IEE POWER AND ENERGY SERIES 47

Series Editors: Professor A. T. Johns
D. F. Warne

Protection of Electricity Distribution Networks - 2nd Edition

Other volumes in this series:

Protection of Electricity Distribution Networks - 2nd Edition

Juan M. Gers and Edward J. Holmes

The Institution of Electrical Engineers

Published by: The Institution of Electrical Engineers, London,
United Kingdom

1004059890 T

British Library Cataloguing in Publication Data

Gers, Juan M.
    Protection of electricity distribution networks – 2nd ed.
    1. Electric power systems – Protection 2. Electric power
    distribution 3. Electric power transmission
    I. Title II. Holmes, E. J. (Edward J), 1928 – III. Institution
    of Electrical Engineers
    621.3'19

ISBN 0 86341 357 9

Typeset in India by Newgen Imaging Systems
Printed in the UK by MPG Books Limited, Bodmin, Cornwall

Contents

Preface and acknowledgements

The quality of electricity supplies is an important factor in the socio-economic development of any area. Approximately 75 per cent of all customer hours lost are owing to faults on the distribution networks, and customers rightly expect a high level of security for their supply. Although this can be achieved by good distribution network design using proven equipment, it is also essential to provide suitable protection schemes and relay settings to ensure that faults are quickly disconnected to minimise outage times and improve the continuity of supplies to customers.

With this in mind this book has been produced as a reference guide for professional engineers and students. It is hoped that the many detailed examples and exercises throughout the book, which the authors have taken from actual case studies in the field, will provide worthwhile material for planning and design engineers and maintenance staff, particularly those engaged in the co-ordination and setting of protection on distribution systems.

The book is based on original material by Mr Gers and translated from the Spanish by Mr Holmes. Subsequently the authors have expanded the text considerably and added much up to date material. It is our view that in the process the continuous dialogue between the co-authors from differing backgrounds and experiences has led to a deeper exploration of the subject matter.

Thanks are due to the University of Valle and the Colombian Institute of Sciences (COLCIENCIAS) for financial help; The British Council for continuing support during many years of institutional exchange; C. Delgado for his valuable comments on the initial manuscripts and our colleagues at GERS Ltd. for their ideas and assistance; F. Pacheco in producing the diagrams; Professor K.L. Lo at Strathclyde University for his guidance, and Stephen and Philip Holmes for sharing so generously their computer expertise. In addition, the authors have been most grateful for the help received from many sources and the permission, readily given, by various organisations to include copyright material which is acknowledged in the text. Finally, we wish to acknowledge the considerable support and understanding that we have received from our wives Pilar and Maggie throughout the four years of work on this book.

J.M. Gers
Cali, Colombia

E.J. Holmes
Stourbridge, England

Preface to 2nd edition

In the six years since this book was first published there have been considerable advances in relay protection design. The development of powerful numerical algorithms and further improvements in digital technology have greatly extended the scope of protection systems. Most of the latest types of relays are now multifunctional devices with control, metering, reporting and alarm functions in addition to their protection capabilities that normally include several types within the same device. They also have very good communication facilities which allow them to work in virtually any automated scheme. Therefore, modern relays now offer better protection coverage and can be programmed to automatically adjust for changes in power system topologies and different operating conditions owing to the multiple setting groups feature incorporated in most of them.

Chapters 3 and 5 have been considerably extended to include more detail on numerical relays. Chapter 12, dealing with protection schemes, has been updated to take account of the new technology available, while the testing procedures covered in the last chapter now include ample reference to numerical protection. The last chapter was also thoroughly updated, particularly in respect of testing procedures applicable to numerical relays.

We have also taken the opportunity to update sections of the original text and have added a new chapter on the processing of alarms since the fast and efficient processing of the many alarms that flow from the power system into control centres has an important bearing on the speed with which system faults are dealt. Our thanks are due to our colleague Professor K.L. Lo of the University of Strathclyde, Glasgow, Scotland, for his help with this material.

Finally, once again we acknowledge the support of our wives Pilar and Maggie during the work on this edition.

J.M. Gers
Weston, Florida
USA
jmgers@gersusa.com

E.J. Holmes
Stourbridge
England
ejholmes@compuserve.com

Chapter 1
Introduction

1.1 General

With the increasing dependence on electricity supplies, in both developing and developed countries, the need to achieve an acceptable level of reliability, quality and safety at an economic price becomes even more important to customers. A further requirement is the safety of the electricity supply. A priority of any supply system is that it has been well designed and properly maintained in order to limit the number of faults that might occur.

Associated with the distribution networks themselves are a number of ancillary systems to assist in meeting the requirements for safety, reliability and quality of supply. The most important of these are the protection systems which are installed to clear faults and limit any damage to distribution equipment. Amongst the principal causes of faults are lightning discharges, the deterioration of insulation, vandalism, and tree branches and animals contacting the electricity circuits. The majority of faults are of a transient nature and can often be cleared with no loss of supply, or just the shortest of interruptions, whereas permanent faults can result in longer outages. In order to avoid damage, suitable and reliable protection should be installed on all circuits and electrical equipment. Protective relays initiate the isolation of faulted sections of the network in order to maintain supplies elsewhere on the system. This then leads to an improved electricity service with better continuity and quality of supply.

A properly co-ordinated protection system is vital to ensure that an electricity distribution network can operate within preset requirements for safety for individual items of equipment, staff and public, and the network overall. Automatic operation is necessary to isolate faults on the networks as quickly as possible in order to minimise damage. The economic costs and the benefits of a protection system must be considered in order to arrive at a suitable balance between the requirements of the scheme and the available financial resources. In addition, minimising the costs of nondistributed energy is receiving increasing attention.

When providing protective devices on any supply network the following basic principles must apply. On the occurrence of a fault or abnormal condition, the

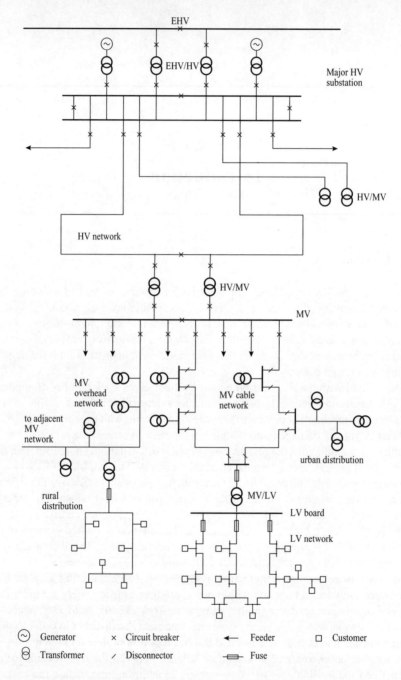

Figure 1.1   EHV/HV/MV/LV network arrangements (reproduced from Electricity Distribution Network Design)

protection system must be capable of detecting it immediately in order to isolate the affected section, thus permitting the rest of the power system to remain in service and limiting the possibility of damage to other equipment. Disconnection of equipment must be restricted to the minimum amount necessary to isolate the fault from the system. The protection must be sensitive enough to operate when a fault occurs under minimum fault conditions, yet be stable enough not to operate when its associated equipment is carrying the maximum rated current, which may be a short-time value. It must also be fast enough to operate in order to clear the fault from the system quickly to minimise damage to system components and be reliable in operation. Back-up protection to cover the possible failure of the main protection is provided in order to improve the reliability of the protection system. While electromechanical relays can still be found in some utilities, the tendency is to replace these by microprocessor and numerical relays, particularly in the more complex protection arrangements.

1.2   Basic principles of electrical systems

The primary aim of any electricity supply system is to meet all customers' demands for energy. Power generation is carried out wherever it achieves the most economic selling cost overall. The transmission system is used to transfer large amounts of energy to major load centres, while distribution systems carry the energy to the furthest customer, using the most appropriate voltage level. Where the transport of very large amounts of power over large distances is involved, an extra high voltage (EHV) system, sometimes termed major or primary transmission, is required. Such systems operate in the 300 kV plus range, typical values being 400, 500 and 765 kV.

High voltage (HV) networks transport large amounts of power within a particular region and are operated either as interconnected systems or discrete groups. Below the transmission system there can be two or three distribution voltage levels to cater for the variety of customers and their demands. In general, the medium voltage (MV) networks and low voltage (LV) networks are operated as radial systems.

Figure 1.1 illustrates the interrelation of the various networks. The HV networks are supplied from EHV/HV substations which themselves are supplied by inter-regional EHV lines. HV/MV transforming substations situated around each HV network supply individual MV networks. The HV and MV networks provide supplies direct to large customers, but the vast majority of customers are connected at low voltage and supplied via MV/LV distribution substations and their associated networks, as shown in Figure 1.2.

1.3   Protection requirements

The protection arrangements for any power system must take into account the following basic principles:

1.   Reliability: the ability of the protection to operate correctly. It has two elements – dependability, which is the certainty of a correct operation on the occurrence of

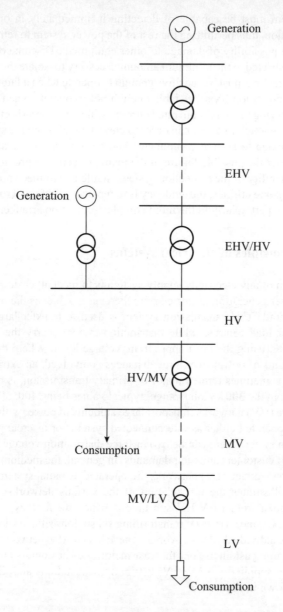

Figure 1.2   Block schematic of transmission and distribution systems (reproduced from Electricity Distribution Network Design)

a fault, and security, which is the ability to avoid incorrect operation during faults.

2. Speed: minimum operating time to clear a fault in order to avoid damage to equipment.

3. Selectivity: maintaining continuity of supply by disconnecting the minimum section of the network necessary to isolate the fault.
4. Cost: maximum protection at the lowest cost possible.

Since it is practically impossible to satisfy all the above-mentioned points simultaneously, inevitably a compromise is required to obtain the optimum protection system.

1.4 Protection zones

The general philosophy for the use of relays is to divide the system into separate zones, which can be individually protected and disconnected on the occurrence of a fault, in order to permit the rest of the system to continue in service wherever possible.

In general a power system can be divided into protection zones – generators, transformers, groups of generator transformers, motors, busbars and lines. Figure 1.3 shows a system with different protection zones. It should be noted that the zones overlap at some points indicating that, if a fault occurs in these overlap areas, more than one set of protection relays should operate. The overlap is obtained by connecting the protection relays to the appropriate current transformers as illustrated in Figure 1.4.

1.5 Primary and back-up protection

All the elements of the power system must be correctly protected so that the relays only operate on the occurrence of a fault. Some relays, designated as unit type protection, operate only for faults within their protection zone. Other relays designated as non-unit protection, are able to detect faults both within a particular zone and also outside it, usually in adjacent zones, and can be used to back up the primary protection as a second line of defence. It is essential that any fault is isolated, even if the associated main protection does not operate. Therefore, wherever possible, every element in the power system should be protected by both primary and back-up relays.

1.5.1 Primary protection

Primary protection should operate every time an element detects a fault on the power system. The protection element covers one or more components of the power system, such as electrical machines, lines and busbars. It is possible for a power system component to have various primary protection devices. However, this does not imply that they all have to operate for the same fault, and it should be noted that the primary protection for one item of system equipment might not necessarily be installed at the same location as the system equipment; in some cases it can be sited in an adjacent substation.

1.5.2 Back-up protection

Back-up protection is installed to operate when, for whatever reason, the primary protection does not work. To achieve this, the back-up protection relay has a sensing

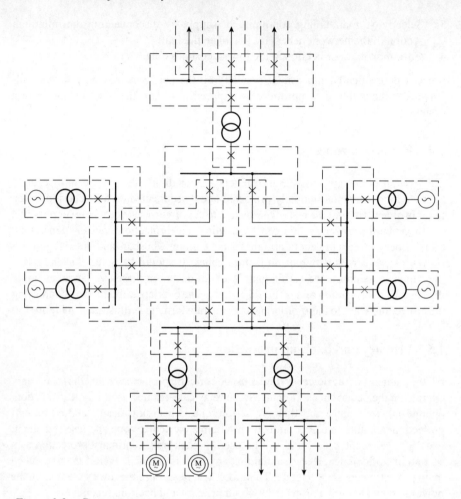

Figure 1.3   Protection zones

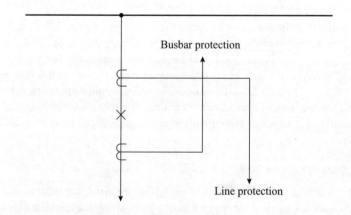

Figure 1.4   Overlap of protection zones

element which may or may not be similar to the primary protection, but which also includes a time-delay facility to slow down the operation of the relay so as to allow time for the primary protection to operate first. One relay can provide back-up protection simultaneously to different pieces of system equipment. Equally the same equipment can have a number of different back-up protection relays and it is quite common for a relay to act as primary protection for one piece of equipment and as back-up for another.

1.6   Directional protection

An important characteristic of some types of protection is their capacity to be able to determine the direction of the flow of power and, by this means, their ability to inhibit opening of the associated switch when the fault current flows in the opposite direction to the setting of the relay. Relays provided with this characteristic are important in protecting mesh networks, or where there are various generation sources, when fault currents can circulate in both directions around the mesh. In these cases, directional protection prevents the unnecessary opening of switchgear and thus improves the security of the electricity supply. On protection schematic diagrams the directional protection is usually represented by an arrow underneath the appropriate symbol, indicating the direction of current flow for relay operation.

Example 1.1 Check on correct operation of protection

Using the power system shown in Figure 1.5, examples are given where there has been incorrect operation of protection and the associated breakers, leading to the operation of back-up protection to isolate the fault from the system, followed by an example

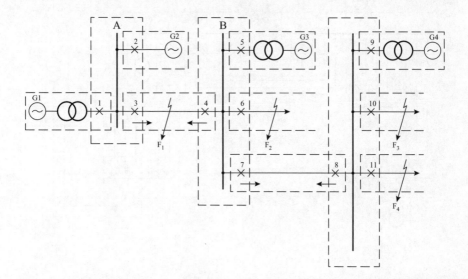

Figure 1.5   Power system for Example 1.1

of correct relay operation, with a final example of unnecessary relay operation. The directional protection is indicated by the arrows below the corresponding breakers.

Table 1.1 shows the breakers that failed to open and those that were tripped by the primary protection and by the back-up protection.

Table 1.1   *Relay/breaker operations for Example 1.1*

Case	Breakers that operated	Breakers that mal-operated	Tripped by primary protection	Tripped by back-up protection
F_1	1, 2, 4	3	4	1, 2
F_2	3, 5, 8	6	–	3, 5, 8
F_3	10	–	10	–
F_4	8, 11	8	11	–

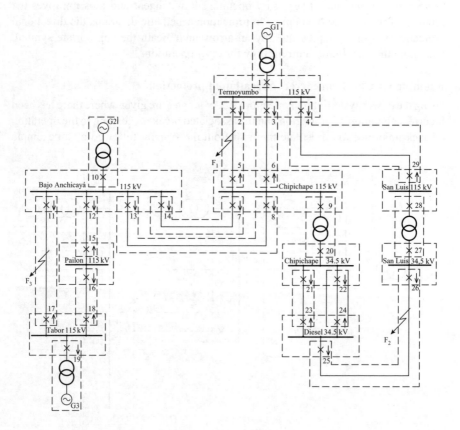

Figure 1.6   *Schematic diagram for Exercise 1.1*

For fault F_1, the protection correctly tripped breaker 4 to open one end of the faulted feeder. With breaker 3 failing to open, breakers 1 and 2 were tripped by back-up protection to stop fault current flowing into the fault from generators G1 and G2. With fault F_2, when breaker 6 failed to operate, the directional protection on breakers 3 and 8 operated to open the incoming feeders from the adjacent busbars, and the back-up protection on breaker 5 tripped to stop G3 feeding into the fault.

Fault F_3 was correctly cleared by the tripping of feeder breaker 10. Fault F_4 was correctly cleared by the operation of breaker 11, so that the tripping of breaker 8 was incorrect. Any fault current flowing along inter-busbar feeder 7-8 before breaker 11 opened would have been from 7 to 8. Relay 8 is directional and operation should not have been initiated for flows from 7 to 8. Thus the first two cases illustrate mal-operation from a dependability point of view, with the last one illustrating mal-operation from a security standpoint.

1.7   Exercise 1.1

For the power system arrangement shown in Figure 1.6, complete Table 1.2, taking into account the operation of the circuit breakers as shown for each fault case. Please note that, as in Example 1.1, some of the circuit breakers that operated may have done so unnecessarily.

Table 1.2   Relay/breaker operations for Exercise 1.1

Case	Breakers that operated	Breakers that mal-operated	Tripped by primary protection	Tripped by back-up protection
F_1	2, 3, 4, 5		2, 5	
F_2	21, 22, 23 24, 27			
F_3	10, 11, 17, 19			

Chapter 2

Calculation of short-circuit currents

The current that flows through an element of a power system is a parameter which can be used to detect faults, given the large increase in current flow when a short-circuit occurs. For this reason a review of the concepts and procedures for calculating fault currents will be made in this chapter, together with some calculations illustrating the methods used. While the use of these short-circuit calculations in relation to protection settings will be considered in detail, it is important to bear in mind that these calculations are also required for other applications, for example calculating the substation earthing grid, the selection of conductor sizes and for the specifications of equipment such as power circuit breakers.

2.1 Modelling for short-circuit current calculations

Electrical faults are characterised by a variation in the magnitude of the short-circuit current due to the effect of the equivalent system impedance at the fault point, which produces a decaying DC component, and the performance of the rotating machinery, which results in a decaying AC component.

2.1.1 Effect of the system impedance

System currents cannot change instantaneously when a fault occurs due to the equivalent system resistances and reactances at the fault point, which result in a decaying DC component. The rate of decay depends on the instantaneous value of the voltage when the fault occurs and the power factor of the system at the fault point. To perform the corresponding calculations, the treatment of electrical faults should be carried out as a function of time, from the start of the event at time $t = 0^+$ until stable conditions are reached, and it is therefore necessary to use differential equations when calculating these currents. In order to illustrate the transient nature of the current, consider an RL circuit as a simplified equivalent of the circuits in electricity distribution networks. This simplification is important because all the system equipment must be modelled

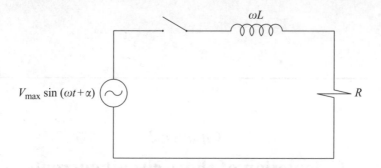

Figure 2.1   RL circuit for transient analysis study

in some way in order to quantify the transient values that can occur during the fault condition.

For the circuit shown in Figure 2.1, the mathematical expression that defines the behaviour of the current is

$$e(t) = L\frac{di}{dt} + Ri(t) \tag{2.1}$$

This is a differential equation with constant coefficients, of which the solution is in two parts:

$$i_a(t) : i_h(t) + i_p(t)$$

where $i_h(t)$ is the solution of the homogeneous equation corresponding to the transient period and $i_p(t)$ is the solution to the particular equation corresponding to the steady-state period.

By the use of differential equation theory, which will not be discussed in detail here, the complete solution can be determined and expressed in the following form:

$$i(t) = \frac{V_{max}}{Z}\left[\sin(\omega t + \alpha - \phi) - \sin(\alpha - \phi)e^{-(R/L)t}\right] \tag{2.2}$$

where

$$Z = \sqrt{(R^2 + \omega^2 L^2)}$$

α = the closing angle, which defines the point on the source sinusoidal voltage when the fault occurs, and

$$\phi = \tan^{-1}(\omega L/R)$$

It can be seen that, in eqn. 2.2, the first term varies sinusoidally and the second term decreases exponentially with a time constant of L/R. The first term corresponds to the AC component, while the second term can be recognised as the DC component of the current having an initial maximum value when $\alpha - \phi = \pm\pi/2$, and zero value when $\alpha = \phi$; see Figure 2.2. It is impossible to predict at what point on the sinusoidal cycle the fault will be applied and therefore what magnitude the DC component will

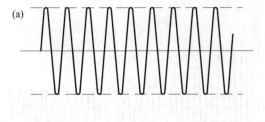

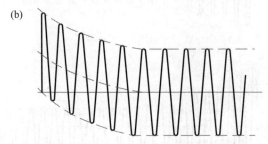

Figure 2.2   Variation of fault current with time: (a) $\alpha - \phi = 0$; (b) $\alpha - \phi = -\pi/2$

reach. If the tripping of the circuit, due to a fault, takes place when the sinusoidal component is at its negative peak, the DC component reaches its theoretical maximum value half a cycle later.

An approximate formula for calculating the effective value of the total asymmetric current, including the AC and DC components, with acceptable accuracy can be used by assuming that these components are in quadrature with the following expression:

$$I_{\text{rms.asym}} = \sqrt{I_{\text{rms}}^2 + I_{\text{DC}}^2} \tag{2.3}$$

2.1.2   Effect of rotating machinery

When a fault occurs close to the terminals of rotating machinery, a decaying AC current is produced, similar in pattern to that flowing when an AC voltage is applied to an RL circuit as discussed in the previous section. Here, the decaying pattern is due to the fact that the magnetic flux in the windings of rotating machinery cannot change instantaneously because of the nature of the magnetic circuits involved. The reduction in current from its value at the onset, due to the gradual decrease in the magnetic flux caused by the reduction of the m.m.f. of the induction current, can be seen in Figure 2.3. This effect is known as armature reaction.

The physical situation that is presented to a generator, and which makes the calculations quite difficult, can be interpreted as a reactance that varies with time. Notwithstanding this, in the majority of practical applications it is possible to take account of the variation of reactance in only three stages without producing significant errors. In Figure 2.4 it will be noted that the variation of current with time, $I(t)$, comes close to the three discrete levels of current, I'', I' and I, the sub-transient, transient and steady state currents respectively. The corresponding values of direct axis reactance

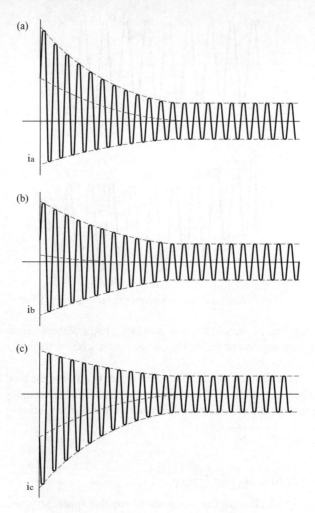

Figure 2.3   Transient short-circuit currents in a synchronous generator: (a) typical
A phase short circuit current; (b) typical B phase short circuit current;
(c) typical C phase short circuit current

are denoted by X_d'', X_d' and X_d, and the typical variation with time for each of these
is illustrated in Figure 2.5.

2.1.3   Types of fault duty

Short-circuit levels vary considerably during a fault, taking into account the rapid
drop of the current due to the armature reaction of the synchronous machines and the
fact that extinction of an electrical arc is never achieved instantaneously. Therefore,
short-circuit currents have to be calculated carefully in order to obtain the correct
value for the respective applications.

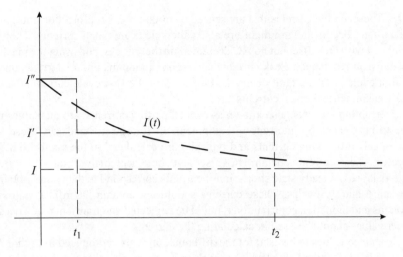

Figure 2.4   Variation of current with time during a fault

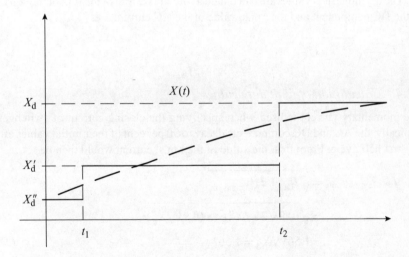

Figure 2.5   Variation of generator reactance with time during a fault

The following paragraphs refer to the short-circuit currents that are specifically used for the selection of interrupting equipment and protection relay settings – the so-called normal duty rating. ANSI/IEEE Standards C37 and IEC 6090 refer to four duty types defined as first cycle or momentary, peak, interrupting or breaking, and time-delayed or steady-state currents.

First cycle currents, also called momentary currents, are the currents present one half of a cycle after fault initiation. In European Standards these values are indicated by I_k''. These are the currents that are sensed by circuit breaker protection equipment when a fault occurs and are therefore also called close and latch currents. They are calculated with DC offset but no AC decrement in the sources, and using the machine sub-transient reactances. Peak currents correspond to the maximum currents during the first cycle after the fault occurs and differ from the first cycle currents that are totally asymmetrical r.m.s. currents.

Interrupting currents, also known as contact parting currents, are the values that have to be cleared by interrupting equipment. In European Standards, these values are called breaking currents and typically are calculated in the range from 3 to 5 cycles. These currents contain DC offset and some decrement of the AC current. Time-delayed or steady state short-circuit currents correspond to the values obtained between 6 and 30 cycles. These currents should not contain DC offset, and synchronous and induction contributions should be neglected and transient reactances or higher values should be used in calculating the currents.

Reactance values to be used for the different duties are reproduced in Figure 2.6, based on IEEE Standard 399-1990. For each case, asymmetrical or symmetrical r.m.s. values can be defined depending on whether the DC component is included or not. The peak values are obtained by multiplying the r.m.s. values by $\sqrt{2}$.

The asymmetrical values are calculated as the square root of the sum of the squares of the DC component and the r.m.s. value of the AC current, i.e.

$$I_{rms} = \sqrt{I_{DC}^2 + I_{AC}^2} \tag{2.4}$$

2.1.4   Calculation of fault duty values

The momentary current is used when specifying the closing current of switchgear. Typically, the AC and DC components decay to 90 per cent of their initial values after the first half cycle. From this, the value of the r.m.s. current would then be

$$I_{rms.asym.closing} = \sqrt{I_{DC}^2 + I_{AC.rms.sym.}^2}$$

$$= \sqrt{(0.9\sqrt{2}V/X_d'')^2 + (0.9V/X_d'')^2}$$

$$= 1.56V/X_d'' = 1.56 I_{rms.sym.} \tag{2.5}$$

Usually a factor of 1.6 is used by manufacturers and in international standards so that, in general, this value should be used when carrying out similar calculations.

The peak value is obtained by arithmetically adding together the AC and DC components. It should be noted that, in this case, the AC component is multiplied by a factor of $\sqrt{2}$. Thus

$$I_{peak} = I_{DC} + I_{AC}$$

$$= (0.9\sqrt{2}V/X_d'') + (0.9\sqrt{2}V/X_d'')$$

$$= 2.55 I_{rms.sym.} \tag{2.6}$$

Duty calculation (see Note 1)	System component	Reactance value for medium- and high-voltage calculations per IEEE Std C37.010-1979 and IEEE Std C37.5-1979	Reactance value for low-voltage calculations (see Note 2)
First cycle (momentary calculations)	**Power company supply**	X_s	X_s
	All turbine generators; all hydro-generators with amortisseur windings; all condensers	$1.0\,X_d''$	$1.0\,X_d''$
	Hydrogenerators without amortisseur windings	$0.75\,X_d''$	$0.75\,X_d'$
	All synchronous motors	$1.0\,X_d''$	$1.0\,X_d''$
	Induction motors		
	Above 1000 hp	$1.0\,X_d''$	$1.0\,X_d''$
	Above 250 hp at 3600 rev/min	$1.0\,X_d''$	$1.0\,X_d''$
	All others, 50 hp and above	$1.2\,X_d''$	$1.2\,X_d''$
	All smaller than 50 hp	$1.67\,X_d''$ (see Note 6)	$1.67\,X_d''$
Interrupting calculations	**Power company supply**	X_s	N/A
	All turbine generators; all hydro-generators with amortisseur windings; all condensers	$1.0\,X_d''$	N/A
	Hydrogenerators without amortisseur windings	$0.75\,X_d'$	N/A
	All synchronous motors	$1.5\,X_d''$	N/A
	Induction motors		
	Above 1000 hp	$1.5\,X_d''$	N/A
	Above 250 hp at 3600 rev/min	$1.5\,X_d''$	N/A
	All others, 50 hp and above	$3.0\,X_d''$	N/A
	All smaller than 50 hp	Neglect	N/A

Notes:

1: First-cycle duty is the momentary (or close-and-latch) duty for medium-/high-voltage equipment and is the interrupting duty for low-voltage equipment.

2: Reactance (X) values to be used for low-voltage breaker duty calculations (see IEEE Std C37.13-1990 and IEEE Std 242-1986).

3: X_d'' of synchronous-rotating machines is the rated-voltage (saturated) direct-axis subtransient reactance.

4: X_d' of synchronous-rotating machines is the rated-voltage (saturated) direct-axis transient reactance.

5: X_d'' of induction motors equals 1 divided by per-unit locked-rotor current at rated voltage.

6: For comprehensive multivoltage system calculations, motors less than 50 hp are represented in medium-/high-voltage short-circuit calculations (see IEEE Std 141-1993, Chapter 4).

Figure 2.6   Reactance values for first cycle and interrupting duty calculations (from IEEE Standard 399-1990; reproduced by permission of the IEEE)

When considering the specification for the switchgear opening current, the so-called r.m.s. value of interrupting current is used in which, again, the AC and DC components are taken into account, and therefore

$$I_{\text{rms.asym.int.}} = \sqrt{I_{\text{DC}}^2 + I_{\text{AC.rms.sym.int.}}^2}$$

Replacing the DC component by its exponential expression gives

$$I_{\text{rms.asym.int.}} = \sqrt{\left(\sqrt{2}I_{\text{rms.sym.int.}}\,e^{-(R/L)t}\right)^2 + I_{\text{rms.sym.int.}}^2}$$

$$= I_{\text{rms.sym.int.}}\sqrt{2e^{-2(R/L)t} + 1} \tag{2.7}$$

The expression $(I_{\text{rms.asym.int.}}/I_{\text{rms.sym.int.}})$ has been drawn for different values of X/R, and for different switchgear contact separation times, in ANSI Standard C37.5-1979. The multiplying factor graphs are reproduced in Figure 2.7.

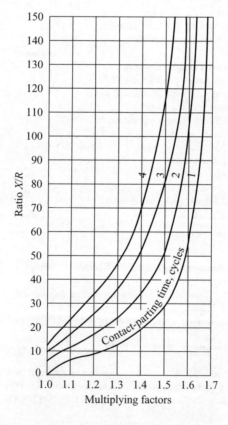

Figure 2.7   Multiplying factors for three-phase and line-to-earth faults (total current rating basis) (from IEEE Standard C37.5-1979; reproduced by permission of the IEEE)

As an illustration of the validity of the curves for any situation, consider a circuit breaker with a total contact separation time of two cycles – one cycle due to the relay and one related to the operation of the circuit breaker mechanism. If the frequency, f, is 60 Hz and the ratio X/R is given as 50, with $t = 2$ cycles $= 0.033$ s, then $(X/R) = (\omega L/R) = 50$. Thus $(L/R) = (50/\omega) = (50/2\pi f) = 0.132$. Therefore

$$\frac{I_{asym}}{I_{sym}} = \sqrt{2\, e^{-(0.033 \times 2)/0.132} + 1} = 1.49$$

as can be seen from Figure 2.7.

In protection co-ordination studies, r.m.s. symmetrical interrupting current values are normally used when setting the time-delay units of the relays. For setting the instantaneous elements, the same values should be used but multiplied by a factor that depends on the application, as will be discussed later on.

2.2 Methods for calculating short-circuit currents

Symmetrical faults, that is three-phase faults and three-phase-to-earth faults, with symmetrical impedances to the fault, leave the electrical system balanced and therefore can be treated by using a single-phase representation. This symmetry is lost during asymmetric faults – line-to-earth, line-to-line, and line-to-line-to-earth – and in these cases a method of analysing the fault that provides a convenient means of dealing with the asymmetry is required. In 1918 a method of symmetrical components was proposed in which an unbalanced system of n related phases could be replaced by a system of n balanced phases which were named the symmetrical components of the original phases. While the method can be applied to any unbalanced polyphase system, the theory is summarised here for the case of an unbalanced three-phase system.

When considering a three-phase system, each vector quantity, voltage or current, is replaced by three components so that a total of nine vectors uniquely represents the values of the three phases. The three system balanced phasors are designated as:

1. Positive-sequence components, which consist of three phasors of equal magnitude, spaced 120° apart, and rotating in the same direction as the phasors in the power system under consideration, i.e. the positive direction.
2. Negative-sequence components, which consist of three phasors of equal magnitude, spaced 120° apart, rotating in the same direction as the positive-sequence phasors but in the reverse sequence.
3. Zero-sequence components, which consist of three phasors equal in magnitude and in phase with each other, rotating in the same direction as the positive-sequence phasors.

With this arrangement, voltage values of any three-phase system, V_a, V_b, and V_c, can be represented thus:

$$V_a = V_{a0} + V_{a1} + V_{a2}$$
$$V_b = V_{b0} + V_{b1} + V_{b2}$$
$$V_c = V_{c0} + V_{c1} + V_{c2}$$

It can be demonstrated that

$$V_b = V_{a0} + a^2 V_{a1} + a V_{a2}$$
$$V_c = V_{a0} + a V_{a1} + a^2 V_{a2}$$

where a is a so-called operator which gives a phase shift of 120° clockwise and a multiplication of unit magnitude i.e. $a = 1\angle 120°$, and a^2 similarly gives a phase shift of 240°, i.e. $a^2 = 1\angle 240°$.

Therefore, the following matrix relationship can be established:

$$\begin{bmatrix} V_a \\ V_b \\ V_c \end{bmatrix} = \begin{bmatrix} 1 & 1 & 1 \\ 1 & a^2 & a \\ 1 & a & a^2 \end{bmatrix} \times \begin{bmatrix} V_{a0} \\ V_{a1} \\ V_{a2} \end{bmatrix}$$

Inverting the matrix of coefficients

$$\begin{bmatrix} V_{a0} \\ V_{a1} \\ V_{a2} \end{bmatrix} = \frac{1}{3} \begin{bmatrix} 1 & 1 & 1 \\ 1 & a & a^2 \\ 1 & a^2 & a \end{bmatrix} \times \begin{bmatrix} V_a \\ V_b \\ V_c \end{bmatrix}$$

From the above matrix it can be deduced that

$$V_{a0} = 1/3(V_a + V_b + V_c)$$
$$V_{a1} = 1/3(V_a + a V_b + a^2 V_c)$$
$$V_{a2} = 1/3(V_a + a^2 V_b + a V_c)$$

The foregoing procedure can also be applied directly to currents, and gives

$$I_a = I_{a0} + I_{a1} + I_{a2}$$
$$I_b = I_{a0} + a^2 I_{a1} + a I_{a2}$$
$$I_c = I_{a0} + a I_{a1} + a^2 I_{a2}$$

Therefore

$$I_{a0} = 1/3(I_a + I_b + I_c)$$
$$I_{a1} = 1/3(I_a + a I_b + a^2 I_c)$$
$$I_{a2} = 1/3(I_a + a^2 I_b + a I_c)$$

In three-phase systems the neutral current is equal to $I_n = (I_a + I_b + I_c)$ and, therefore, $I_n = 3 I_{a0}$. By way of illustration, a three-phase unbalanced system is shown in Figure 2.8 together with the associated symmetrical components.

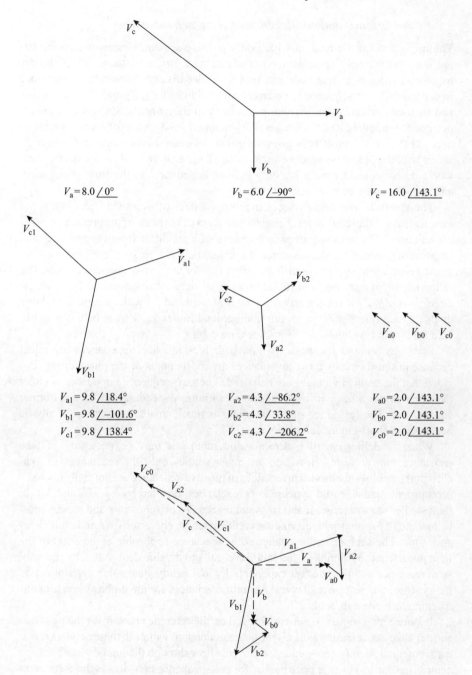

$V_a = 8.0 \, \underline{/0°}$ $\qquad$ $V_b = 6.0 \, \underline{/-90°}$ $\qquad$ $V_c = 16.0 \, \underline{/143.1°}$

$V_{a1} = 9.8 \, \underline{/18.4°}$ $\qquad$ $V_{a2} = 4.3 \, \underline{/-86.2°}$ $\qquad$ $V_{a0} = 2.0 \, \underline{/143.1°}$

$V_{b1} = 9.8 \, \underline{/-101.6°}$ $\qquad$ $V_{b2} = 4.3 \, \underline{/33.8°}$ $\qquad$ $V_{b0} = 2.0 \, \underline{/143.1°}$

$V_{c1} = 9.8 \, \underline{/138.4°}$ $\qquad$ $V_{c2} = 4.3 \, \underline{/-206.2°}$ $\qquad$ $V_{c0} = 2.0 \, \underline{/143.1°}$

Figure 2.8   Symmetrical components of an unbalanced three-phase system

2.2.1   *Importance and construction of sequence networks*

The impedance of a circuit in which only positive-sequence currents are circulating is called the positive-sequence impedance and, similarly, those in which only negative- and zero-sequence currents flow are called the negative- and zero-sequence impedances. These sequence impedances are designated Z_1, Z_2 and Z_0 respectively and are used in calculations involving symmetrical components. Since generators are designed to supply balanced voltages, the generated voltages are of positive sequence only. Therefore the positive-sequence network is composed of an e.m.f. source in series with the positive-sequence impedance. The negative- and zero-sequence networks do not contain e.m.f.s but only include impedances to the flow of negative- and zero-sequence currents respectively.

The positive- and negative-sequence impedances of overhead line circuits are identical, as are those of cables, being independent of the phase if the applied voltages are balanced. The zero-sequence impedances of lines differ from the positive- and negative-sequence impedances since the magnetic field creating the positive- and negative-sequence currents is different from that for the zero-sequence currents. The following ratios may be used in the absence of detailed information. For a single circuit line, $Z_0/Z_1 = 2$ when no earth wire is present and 3.5 with an earth wire. For a double circuit line $Z_0/Z_1 = 5.5$. For underground cables Z_0/Z_1 can be taken as 1 to 1.25 for single core, and 3 to 5 for three-core cables.

For transformers, the positive- and negative-sequence impedances are equal because in static circuits these impedances are independent of the phase order, provided that the applied voltages are balanced. The zero-sequence impedance is either the same as the other two impedances, or infinite, depending on the transformer connections. The resistance of the windings is much smaller and can generally be neglected in short-circuit calculations.

When modelling small generators and motors it may be necessary to take resistance into account. However, for most studies only the reactances of synchronous machines are used. Three values of positive-reactance are normally quoted – subtransient, transient and synchronous reactances, denoted by X_d'', X_d' and X_d. In fault studies the subtransient and transient reactances of generators and motors must be included as appropriate, depending on the machine characteristics and fault clearance time. The subtransient reactance is the reactance applicable at the onset of the fault occurrence. Within 0.1 s the fault level falls to a value determined by the transient reactance and then decays exponentially to a steady-state value determined by the synchronous reactance. Typical per-unit reactances for three-phase synchronous machines are given in Table 2.1.

In connecting sequence networks together, the reference busbar for the positive- and negative-sequence networks is the generator neutral which, in these networks, is at earth potential so only zero-sequence currents flow through the impedances between neutral and earth. The reference busbar for zero-sequence networks is the earth point of the generator. The current that flows in the impedance Z_n between the neutral and earth is three times the zero-sequence current. Figure 2.9 illustrates the sequence networks for a generator. The zero-sequence network carries only zero-sequence current in one phase, which has an impedance of $Z_0 = 3Z_n + Z_{e0}$.

Table 2.1   Typical per-unit reactances for three-phase synchronous machines

Type of machine		X''_d	X'_d	X_d	X_2	X_0
Turbine	2 pole	0.09	0.15	1.20	0.09	0.03
Generator	4 pole	0.14	0.22	1.70	0.14	0.07
Salient pole	with dampers	0.20	0.30	1.25	0.20	0.18
Generator	without dampers	0.28	0.30	1.20	0.35	0.12

The voltage and current components for each phase are obtained from the equations given for the sequence networks. The equations for the components of voltage, corresponding to the a phase of the system, are obtained from the point a on phase a relative to the reference busbar, and can be deduced from Figure 2.9 as follows:

$$V_{a1} = E_a - I_{a1}Z_1$$

$$V_{a2} = -I_{a2}Z_2$$

$$V_{a0} = -I_{a0}Z_0$$

where: E_a = no load voltage to earth of the positive-sequence network; Z_1 = positive-sequence impedance of the generator; Z_2 = negative-sequence impedance of the generator; Z_0 = zero-sequence impedance of the generator (Z_{g0}) plus three times the impedance to earth.

The above equations can be applied to any generator that carries unbalanced currents and are the starting point for calculations for any type of fault. The same approach can be used with equivalent power systems or to loaded generators, E_a then being the voltage behind the reactance before the fault occurs.

2.2.2   Calculation of asymmetrical faults using symmetrical components

The positive-, negative- and zero-sequence networks, carrying currents I_1, I_2 and I_0 respectively, are connected together in a particular arrangement to represent a given unbalanced fault condition. Consequently, in order to calculate fault levels using the method of symmetrical components, it is essential to determine the individual sequence impedances and combine these to make up the correct sequence networks. Then, for each type of fault, the appropriate combination of sequence networks is formed in order to obtain the relationships between fault currents and voltages.

Line-to-earth fault

The conditions for a solid fault from line a to earth are represented by the equations $I_b = 0$, $I_c = 0$ and $V_a = 0$. As in the previous equations, it can easily be deduced that $I_{a1} = I_{a2} = I_{a0} = E_a/(Z_1 + Z_2 + Z_0)$. Therefore, the sequence networks will be connected in series, as indicated in Figure 2.10a. The current and voltage conditions are the same when considering an open-circuit fault in phases b and c, and thus the treatment and connection of the sequence networks will be similar.

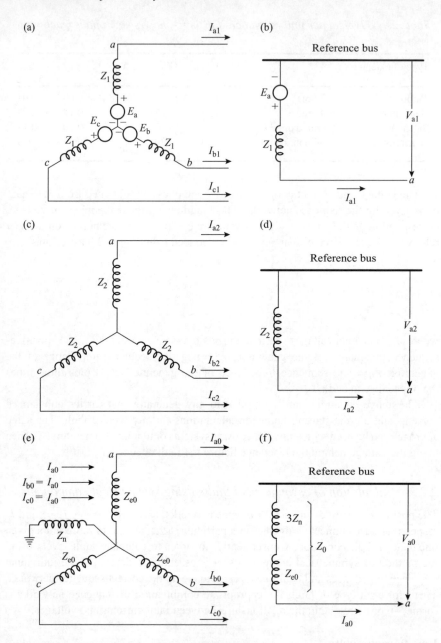

*Figure 2.9   Equivalent sequence networks and current flows for a synchronous
generator:* (a) positive-sequence current; (b) positive-sequence network;
(c) negative-sequence current; (d) negative-sequence network; (e) zero-
sequence current; (f) zero-sequence network

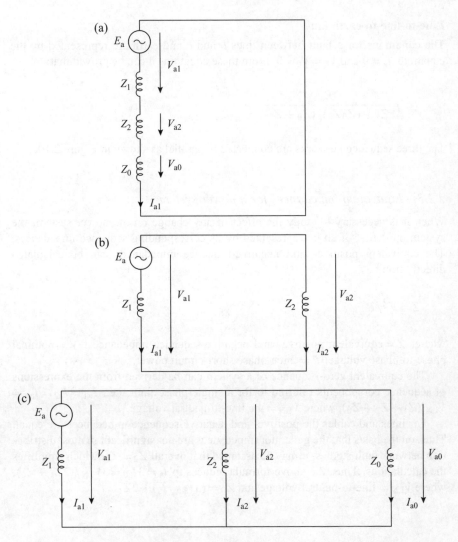

Figure 2.10   Connection of sequence networks for asymmetrical faults: (a) phase-to-earth fault; (b) phase-to-phase fault; (c) double phase-to-earth fault

Line-to-line fault

The conditions for a solid fault between lines b and c are represented by the equations $I_a = 0$, $I_b = -I_c$ and $V_b = V_c$. Equally it can be shown that $I_{a0} = 0$ and $I_{a1} = E_a/(Z_1 + Z_2) = -I_{a2}$. For this case, with no zero-sequence current, the zero-sequence network is not involved and the overall sequence network is composed of the positive- and negative-sequence networks in parallel as indicated in Figure 2.10b.

Line-to-line-to-earth fault

The conditions for a fault between lines b and c and earth are represented by the equations $I_a = 0$ and $V_b = V_c = 0$. From these equations it can be proved that:

$$I_{a1} = \frac{E_a}{Z_1 + (Z_0 Z_2)/(Z_0 + Z_2)}$$

The three sequence networks are connected in parallel as shown in Figure 2.10c.

2.2.3   Equivalent impedances for a power system

When it is necessary to study the effect of any change on the power system, the system must first of all be represented by its corresponding sequence impedances. The equivalent positive- and negative-sequence impedances can be calculated directly from:

$$Z = V^2/P$$

where: Z = equivalent positive- and negative-sequence impedances; V = nominal phase-to-phase voltage; P = three-phase short-circuit power.

The equivalent zero-sequence of a system can be derived from the expressions of sequence components referred to for a single-phase fault, i.e., $I_{a1} = I_{a2} = I_{a0} = V_{LN}/(Z_1 + Z_2 + Z_0)$, where V_{LN} = the line-to-neutral voltage.

For lines and cables the positive- and negative-sequence impedances are equal. Thus, on the basis that the generator impedances are not significant in most distribution network fault studies, it may be assumed that overall $Z_2 = Z_1$, which simplifies the calculations. Thus, the above formula reduces to $I_a = 3 I_{a0} = 3 V_{LN}/(2Z_1 + Z_0)$, where V_{LN} = line-to-neutral voltage and $Z_0 = (3 V_{LN}/I_a) - 2Z_1$.

2.3   Supplying the current and voltage signals to protection systems

In the presence of a fault the current transformers (CTs) circulate current proportional to the fault current to the protection equipment without distinguishing between the vectorial magnitudes of the sequence components. Therefore, in the majority of cases, the relays operate on the basis of the corresponding values of fault current and/or voltages, regardless of the values of the sequence components. It is very important to emphasise that, given this, the advantage of using symmetrical components is that they facilitate the calculation of fault levels even though the relays in the majority of cases do not distinguish between the various values of the symmetrical components.

In Figure 2.11 the positive- and negative-sequence values of current and voltage for different faults are shown together with the summated values of current and

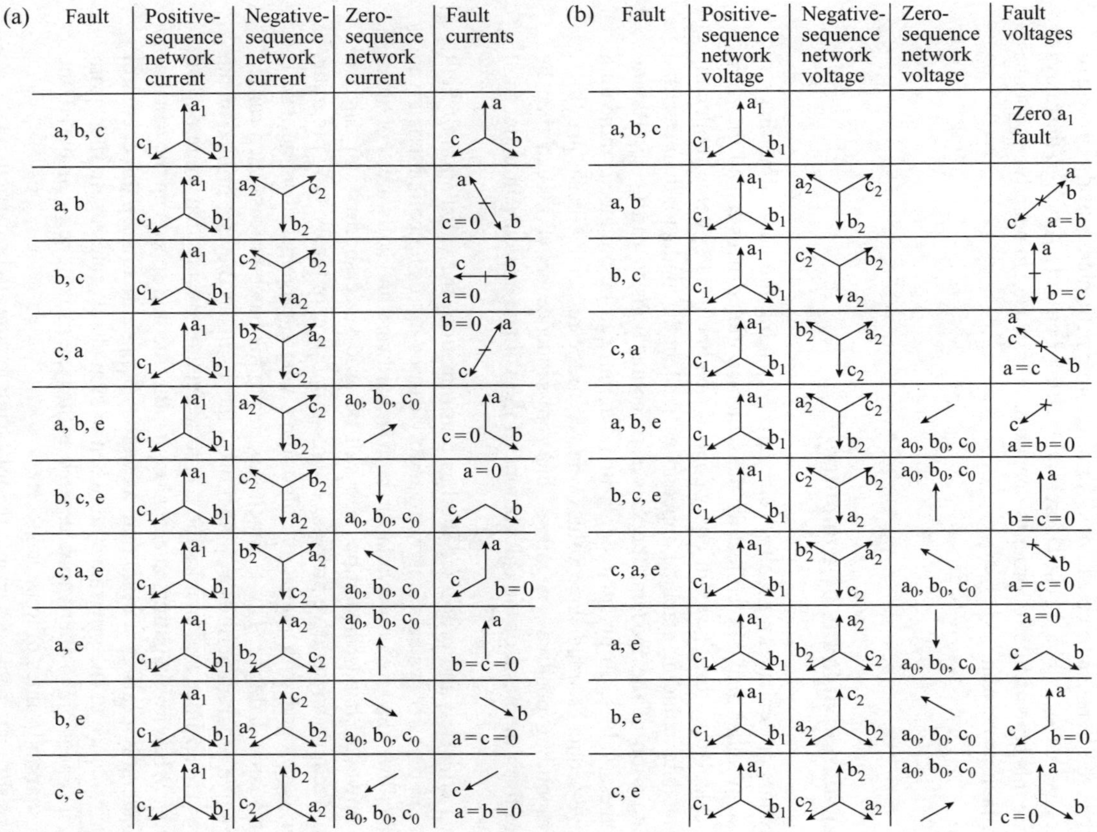

Figure 2.11 Currents and voltages for various types of faults: (a) sequence currents for different types of fault; (b) sequence voltages for different types of fault

voltage. Relays usually only operate using the summated values in the right-hand columns. However, relays are available that can operate with specific values of some of the sequence components. In these cases there must be methods for obtaining these components, and this is achieved by using filters that produce the mathematical operations of the resultant equations to resolve the matrix for voltages and for currents. Although these filters can be constructed for electromagnetic elements, the growth of electronics has led to their being used increasingly in logic circuits. Amongst the relays that require this type of filter in order to operate are those used in negative-sequence and earth-fault protection.

2.4   Calculation of faults by computer

The procedure for calculating fault levels starts by taking the single-line diagram of the system under analysis; collecting the sequence impedances for all the components; calculating the Thevenin equivalent of neighbouring systems; collecting background data including machine impedances, the length, conductor diameter, and the configuration of the feeders, the values of connections to earth, etc. Having obtained these values, an updated single-line diagram can be produced, indicating the positive-sequence impedance values referred to the respective base quantities. The corresponding positive-, negative- and zero-sequence networks can then be built up to form the basis of the calculations of the voltage and current under fault conditions.

Having collected and processed the basic information, the calculation of fault levels for large power systems is now invariably carried out using computers, given the vast facilities that are now available, both in hardware and software. For small systems, however, hand calculations can still be used since short-circuit calculations do not require an iterative process. A large proportion of the existing programs have been developed with interactive algorithms whose principal characteristic is the man-machine dialogue. This is much superior to the batch process used earlier. The interactive program permits examination of the results as they are printed on the screen or via the printer, and enables the user to select those results that are important in the study. When investigating electrical faults, this method speeds up the calculations considerably bearing in mind that, as well as being able to keep direct control on the performance of the program, (i.e. when to stop/check, interrogate files, print out, etc.), it is possible to alter variables such as the type of fault being analysed, the faulted busbar and values of impedance, etc.

Modern software packages enable the following features to be carried out:

- Calculations not only for the standard fault types, i.e. three-phase, line-to-line, line-to-line-to-earth, line-to-earth, but also for faults such as those between systems of different voltages;

- Indicate the fault contributions, (sequence and phase values), from the different elements whether or not they are associated with the faulted node;
- Include pre-fault values;
- Calculate the different duties associated with a fault and handle IEEE and IEC Standards, which can be slightly different especially in the pre-fault voltage level;
- Calculate simultaneous faults;
- Calculate faults along different line lengths.

Chapter 3

Classification and function of relays

A protection relay is a device that senses any change in the signal it is receiving, usually from a current and/or voltage source. If the magnitude of the incoming signal is outside a pre-set value the relay will carry out a specific operation, generally to close or open electrical contacts to initiate some further operation, for example the tripping of a circuit breaker.

3.1 Classification

Protection relays can be classified in accordance with their construction, the incoming signal and function.

3.1.1 Construction

- electromechanical
- solid state
- microprocessor
- numerical
- non-electric (thermal, pressure, etc.)

3.1.2 Incoming signal

- current
- voltage
- power
- frequency
- temperature
- pressure
- speed
- others

3.1.3   Function

- overcurrent
- directional overcurrent
- distance
- overvoltage
- differential
- reverse power
- others

3.1.4   International identification of electrical devices

The international classification for the more common relays, which is used in the following chapters, is given below:

21	distance relay
24	volts/hertz
25	synchronising or synchronism-check device
26	thermal device
27	undervoltage relay
32	reverse-power relay
37	under-current or under-power relay
40	relay for field excitation
41	field circuit breaker
43	manual transfer or selector device
46	negative-sequence current relay
47	negative-sequence voltage relay
49	thermal relay
50	instantaneous overcurrent relay
51	time-delay overcurrent relay
52	circuit breaker
55	power factor relay
59	overvoltage relay
60	voltage or current balance relay
62	time-delay relay
63	pressure relay, for flow or level of liquid or gases
64	earth protection relay
67	directional overcurrent relay
68	blocking relay
74	alarm relay
78	out-of-step relay
79	reclosing relay
81	frequency relay
85	carrier or pilot-wire receiver relay
86	lockout relay
87	differential relay
94	auxiliary tripping relay

In some cases a letter is added to the number associated with the protection in order to specify its place of location, for example G for generator, T for transformer, etc. Non-electric relays are outside the scope of this book and therefore are not referred to.

3.2   Electromechanical relays

These relays are constructed with electrical, magnetic and mechanical components and have an operating coil and various contacts, and are very robust and reliable. They are also referred to as electromagnetic relays due to their magnetic components. Their construction characteristics can be classified in three groups, as detailed below.

3.2.1   Attraction relays

Attraction relays can be supplied by AC or DC, and operate by the movement of a piece of metal when it is attracted by the magnetic field produced by a coil. There are two main types of relay in this class. The attracted armature type, which is shown in Figure 3.1, consists of a bar or plate of metal that pivots when it is attracted towards the coil. The armature carries the moving part of the contact which is closed or opened, according to the design, when the armature is attracted to the coil. The other type is the piston or solenoid type relay, illustrated in Figure 3.2, in which a bar or piston is attracted axially within the field of the solenoid. In this case, the piston also carries the operating contacts.

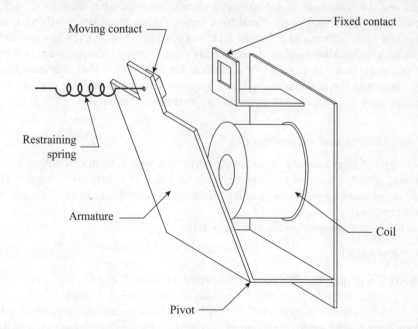

Figure 3.1   Armature-type relay

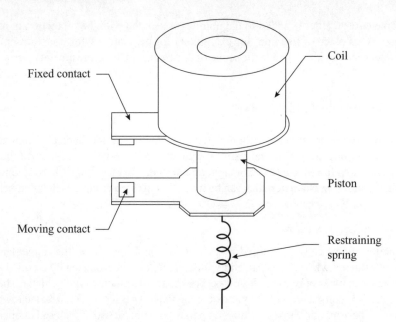

Figure 3.2   Solenoid-type relay

It can be shown that the force of attraction is equal to $K_1 I^2 - K_2$, where K_1 depends upon the number of turns on the operating solenoid, the air gap, the effective area and the reluctance of the magnetic circuit, amongst other factors. K_2 is the restraining force, usually produced by a spring. When the relay is balanced, the resultant force is zero and therefore $K_1 I^2 = K_2$, so that $I = \sqrt{(K_1/K_2)} = \text{constant}$. In order to control the value at which the relay starts to operate, the restraining tension of the spring or the resistance of the solenoid circuit can be varied, thus modifying the restricting force. Attraction relays effectively have no time delay and, for that reason, are widely used when instantaneous operation is required.

3.2.2   Relays with moveable coils

This type of relay consists of a rotating movement with a small coil suspended or pivoted with the freedom to rotate between the poles of a permanent magnet. The coil is restrained by two springs which also serve as connections to carry the current to the coil.

The torque produced in the coil is given by

$$T = BlaNi$$

where: $T = $ torque; $B = $ flux density; $l = $ length of the coil; $a = $ diameter of the coil; $N = $ number of turns on the coil; $i = $ current flowing through the coil.

From the above equation it will be noted that the torque developed is proportional to the current. The speed of movement is controlled by the damping action which is

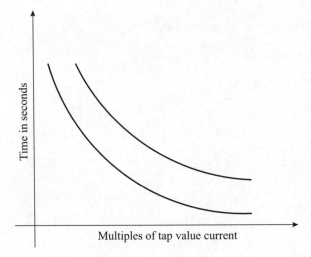

Figure 3.3   Inverse time characteristic

proportional to the torque. It thus follows that the relay has an inverse time characteristic similar to that illustrated in Figure 3.3. The relay can be designed so that the coil makes a large angular movement, for example 80°.

3.2.3   Induction relays

An induction relay works only with alternating current. It consists of an electromagnetic system which operates on a moving conductor, generally in the form of a disc or cup, and functions through the interaction of electromagnetic fluxes with the parasitic Foucalt currents that are induced in the rotor by these fluxes. These two fluxes, which are mutually displaced both in angle and in position, produce a torque that can be expressed by $T = K_1 \Phi_1 \Phi_2 \sin \Theta$, where Φ_1 and Φ_2 are the interacting fluxes and Θ is the phase angle between Φ_1 and Φ_2. It should be noted that the torque is a maximum when the fluxes are out of phase by 90° and zero when they are in phase.

It can be shown that $\Phi_1 = \Phi_1 \sin \omega t$, and $\Phi_2 = \Phi_2 \sin(\omega t + \Theta)$, where Θ is the angle by which Θ_2 leads Φ_1.

Then,

$$i_{\Phi_1} \propto \frac{d\Phi_1}{dt} \propto \Phi_1 \cos \omega t$$

and

$$i_{\Phi_2} \propto \frac{d\Phi_2}{dt} \propto \Phi_2 \cos(\omega t + \Theta)$$

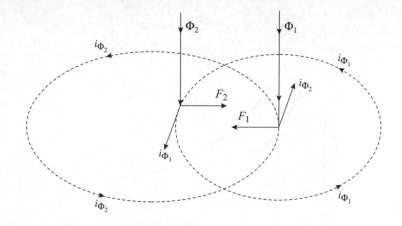

Figure 3.4   Electromagnetic forces in induction relays

Figure 3.4 shows the interrelationship between the currents and the opposing forces.

Thus:

$$F = (F_2 - F_1) \propto (\Phi_2 i_{\Phi_1} - \Phi_1 i_{\Phi_2})$$
$$F \propto \Phi_2 \sin(\omega t + \Theta)\Phi_1 \cos \omega t - \Phi_1 \sin \omega t \, \Phi_2 \cos(\omega t + \Theta)$$
$$F \propto \Phi_1 \Phi_2 [\sin(\omega t + \Theta) \cos \omega t - \sin \omega t \cos(\omega t + \Theta)]$$
$$F \propto \Phi_1 \Phi_2 [\sin\{(\omega t + \Theta) - \omega t\}]$$
$$F \propto \Phi_1 \Phi_2 \sin \Theta \propto T$$

Induction relays can be grouped into three classes as set out below.

(i)   Shaded pole relay

In this case a portion of the electromagnetic section is short-circuited by means of a copper ring or coil. This creates a flux in the area influenced by the short-circuited section (the so-called shaded section) which lags the flux in the non-shaded section (see Figure 3.5).

(ii)   Wattmetric type relay

In its more common form, this type of relay uses an arrangement of coils above and below the disc with the upper and lower coils fed by different values or, in some cases, with just one supply for the top coil, which induces an out-of-phase flux in the lower coil because of the air gap. Figure 3.6 illustrates a typical arrangement.

(iii)   Cup type relay

This type of relay has a cylinder similar to a cup which can rotate in the annular air gap between the poles of the coils, and has a fixed central core (see Figure 3.7).

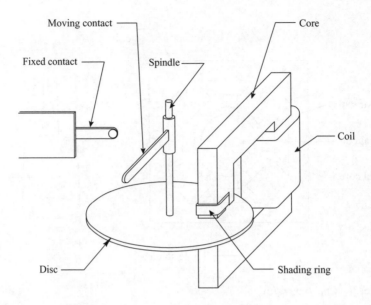

Figure 3.5 Shaded-pole relay

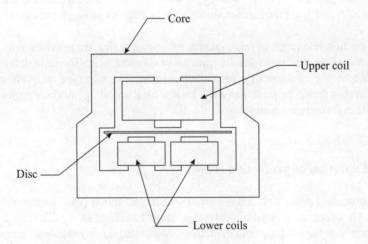

Figure 3.6 Wattmetric-type relay

The operation of this relay is very similar to that of an induction motor with salient poles for the windings of the stator. Configurations with four or eight poles spaced symmetrically around the circumference of the cup are often used. The movement of the cylinder is limited to a small amount by the contact and the stops. A special spring provides the restraining torque. The torque is a function of the product of

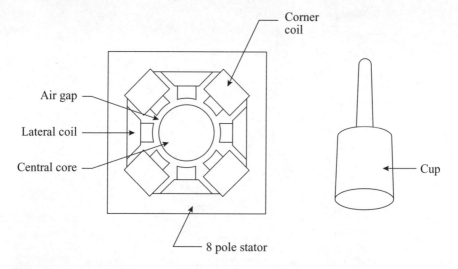

Figure 3.7   Cup-type relay

the two currents through the coils and the cosine of the angle between them. The torque equation is $T = [K I_1 I_2 \cos(\Theta_{12} - \Phi) - K_s]$, where K, K_s, and Φ are design constants, I_1 and I_2 are the currents through the two coils, and Θ_{12} is the angle between I_1 and I_2.

In the first two types of relay mentioned above, which are provided with a disc, the inertia of the disc provides the time delay characteristic. The time delay can be increased by the addition of a permanent magnet. The cup type relay has a small inertia and is therefore principally used when high-speed operation is required, for example in instantaneous units.

3.3   Evolution of protection relays

The evolution of protection relays started with the attraction type of relay referred to earlier. However, the design of protection relays has changed significantly over the past years with the advancement in microprocessor and signal processing technology. As electronic technology progressed, electromechanical relays were superseded in the 1960s by electronic or static designs using transistors and similar types of electronic elements. The integrated circuit (IC) enabled static designs to be further extended and improved in the 1970s. Following the development of the microprocessor, basic programmable microprocessor or micro-controlled multifunction protection relays first started to appear in the early 1980s. Subsequently in the 1990s microprocessor technology, along with the improvements in mathematical algorithms, spurred the development of the so-called numerical relays which are extremely popular for their multifunctional capabilities, low prices and reliability.

3.4   Numerical protection

3.4.1   General

Numerical protection relays operate on the basis of sampling inputs and controlling outputs to protect or control the monitored system. System currents and/or voltages, for example, are not monitored on a continuous basis but, like all other quantities, are sampled one at a time. After acquiring samples of the input waveforms, calculations are performed to convert the incremental sampled values into a final value that represents the associated input quantity based on a defined algorithm. Once the final value of an input quantity can be established, the appropriate comparison to a setting, or reference value, or some other action, can be taken as necessary by the protection relay. Depending upon the algorithm used, and other system design or protection requirements, the final value may be calculated many times within a single sampling cycle, or only once over many cycles.

Most numerical relays are multifunctional and can be regarded as intelligent electronic devices (IEDs). The proper handling of all the features requires a flexible programmable logic platform for the user to apply the available functions with complete flexibility and be able to customise the protection to meet the requirements of the protected power system. Programmable I/O, extensive communication features and an advanced human-machine interface (HMI), which is normally built into most relays, provide easy access to the available features.

3.4.2   Characteristics of numerical relays

Numerical relays are technically superior to the conventional types of relays described earlier in this chapter. Their general characteristics are:

- Reliability: incorrect operations are less likely with numerical relays.
- Self-diagnosis: numerical relays have the ability to conduct continuous self-diagnosis in the form of watchdog circuitry, which includes memory checks and analogue input module tests. In case of failure, normally the relays either lock out or attempt a recovery, depending on the disturbance detected.
- Event and disturbance records: these relays can produce records of events whenever there is a protection function operation, the energising of a status input, or any hardware failure. In addition, disturbance records can be generated in a number of analogue channels, together with all the status input and relay output information.
- Integration of digital systems: the present technology now includes many other tasks at one substation, such as communications, measurement and control. These functions can be integrated into one digital system so that a substation can be operated in a more rapid and reliable manner. Fibre optics are now being used to provide communication links between various system elements to avoid the interference problems that can occur when using metallic conductors.
- Adaptive protection: with the programming and communication capacity of digital systems, the numerical relay can provide adaptive protection. This feature

enables the relay setting to be changed depending on the operating conditions of the network, thus guaranteeing suitable relay settings for the real-time situation by not using a setting based on the most critical system arrangement, which sometimes does not provide the most appropriate solution. The algorithms for relay settings are usually in low-level languages because of the need for a short time response which is not obtained with high-level languages such as Pascal or Fortran.

3.4.3   Typical architectures of numerical relays

Numerical relays are made up from modules with well-defined functions. Figure 3.8 shows a block diagram for numerical relays using typical modules.

The main modules are as follows:

- Microprocessor: responsible for processing the protection algorithms. It includes the memory module which is made up from two memory components:
    - RAM (random access memory), which has various functions, including retaining the incoming data that is input to the processor and is necessary for storing information during the compilation of the protection algorithm.
    - ROM (read only memory) or PROM (programmable ROM), which are used for storing programs permanently.
- Input module: the analogue signals from the substation are captured and sent to the microprocessor and the module typically contains the following

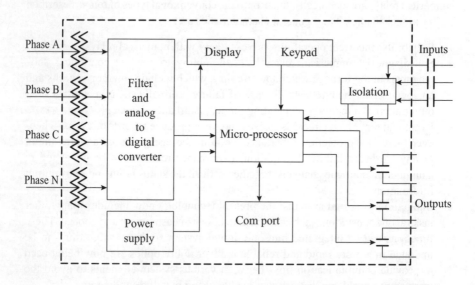

Figure 3.8   General arrangement of numerical relays (reproduced by permission of Basler Electric)

elements:
- analogical filters, which are active low-bandpass filters that eliminate any background noise that has been induced in the line;
- signal conditioner, which converts the signal from the CTs into a normalised DC signal;
- analogue digital convertor, which converts the normalised DC signal into a binary number that is then sent directly to the microprocessor or to a communications buffer.

- Output module: conditions the microprocessor response signals and sends them to the external elements that they control. It is made up of a digital output which generates a pulse as a response signal, and a signal conditioner which amplifies and isolates the pulse.
- Communication module: contains series and parallel ports to permit the interconnection of the protection relays with the control and communications systems of the substation.

3.4.4   Standard functions of numerical relays

Numerical multifunction relays are similar in nature to a panel of single-function protection relays. Both must be wired together with ancillary devices to operate as a complete protection and control system. In the single-function and electromagnetic environment, diagrams provide information on wiring protection elements, switches, meters, and indicator lights. In the digital multifunction environment the process of wiring individual protection or control elements is replaced by entering logic settings. The process of creating a logic scheme is the digital equivalent of wiring a panel. It integrates the multifunction protection, control, and input/output elements into a unique protection and control system.

(i)   Protection

Protection functions of numerical relays may typically include one or more of the following: directional/non-directional three-phase overcurrent; directional/non-directional earth-fault overcurrent; negative-sequence overcurrent; directional power; over-excitation; over- and under-voltage; over- and under-frequency; distance; field loss; differential; breaker failure; automatic reclosing; breaker monitoring and automatic reclosing.

(ii)   Measurement

Numerical relays normally incorporate outstanding measurement functions. Three-phase currents and voltages are digitally sampled and the fundamental is extracted using a discrete Fourier transform (DFT) algorithm. Metering functions include voltage, current, frequency, power factor, apparent power, reactive power and true power. Metered values are viewed through any communication port using serial commands or at the front panel HMI if available.

(iii)   Control

Most numerical relays incorporate at least one virtual breaker control switch and several virtual switches which can be accessed locally from the HMI or remotely from the communications port. The virtual breaker control switch permits the tripping and closing of a selected breaker. The virtual switches can be used to trip and close additional switches and breakers, or enable and disable certain functions.

(iv)   Communication

In numerical relaying, relay and power system information can be retrieved from a remote location using the ASCII command interface which can also be used to enter settings, retrieve reports and metering information, and perform control operations. A communication port on the relay front panel provides a temporary local interface for communication. Communication ports on the rear panel provide a permanent communication interface.

Panel communication ports can be connected to computers, terminals, serial printers, modems, and logic intermediate communication/control interfaces such as RS-232 serial multiplexors. Most numerical relay communication protocols support ASCII and binary data transmissions. ASCII data is used to send and receive human readable data and commands. Binary data is used for computer communication and transmission of raw oscillographic fault data if available. In most numerical relays, at least one of the following protocols is available – Modbus, DNP, Courier, IEC 608750-5 and MMS/UCA2.

(v)   Reporting and alarms

The fault reporting functions provide means of recording and reporting information about faults that have been detected by the relay. The most basic fault reporting functions provided by a protection relay are the signalling or visual flags which indicate the type of fault, usually referred to as targets. In addition, the numerical relay can provide many advanced fault reporting features. These include fault summary reports, sequence of events recorder reports, and oscillographic records. Above all, it is essential to be able to download information to COMTRADE files, which can be opened by numerous programs.

In numerical relays, the settings are introduced as logic equations and trip expressions are used by the fault reporting function to start logging targets for an event and to record the fault current magnitudes at the time of trip. The HMI uses the trip expression to display the respective target by means of the associated LED. The breaker monitoring function also uses the trip expression to start counting the breaker operating time.

Pick-up expressions are used by the fault reporting function to time-stamp the fault summary record, time the length of the fault from pick-up to drop-out (fault clearance time) and to control the recording of oscillograph data. The HMI also uses the pick-up expression to control the flashing of a trip LED. A pick-up expression is also used by the setting group selection function to prevent a setting group change during a fault.

3.5   Supplies to the relay circuits

Protection relays are usually designed for either alternating current or direct current circuits, the two types generally being totally independent of each other. Voltage and/or current signals are taken from measurement transformers to feed the AC circuits. In the case of the so-called primary relays the connection is made directly to the supply system. The AC signals feed the control circuits of the relays which then determine whether or not fault conditions exist. Various alarm and control signals (for example, those used to open switches) are carried along DC circuits. These circuits normally obtain their supply from banks of batteries so that faults on the AC system do not affect the operation of the switchgear mechanisms.

Chapter 4

Current and voltage transformers

Current or voltage instrument transformers are necessary to isolate the protection, control and measurement equipment from the high voltages of a power system, and for supplying the equipment with the appropriate values of current and voltage – generally these are 1 A or 5 A for the current coils, and 120 V for the voltage coils. The behaviour of current and voltage transformers during and after the occurrence of a fault is critical in electrical protection since errors in the signal from a transformer can cause mal-operation of the relays. In addition, factors such as the transient period and saturation must be taken into account when selecting the appropriate transformer. When only voltage or current magnitudes are required to operate a relay then the relative direction of the current flow in the transformer windings is not important. However, the polarity must be kept in mind when the relays compare the sum or difference of the currents.

4.1 Voltage transformers

With voltage transformers (VTs) it is essential that the voltage from the secondary winding should be as near as possible proportional to the primary voltage. In order to achieve this, VTs are designed in such a way that the voltage drops in the windings are small and the flux density in the core is well below the saturation value so that the magnetisation current is small; in this way a magnetisation impedance is obtained that is practically constant over the required voltage range. The secondary voltage of a VT is usually 115 or 120 V with corresponding line-to-neutral values. The majority of protection relays have nominal voltages of 120 or 69 V, depending on whether their connection is line-to-line or line-to-neutral.

4.1.1 Equivalent circuit

VTs can be considered as small power transformers so that their equivalent circuit is the same as for power transformers, as shown in Figure 4.1a. The magnetisation

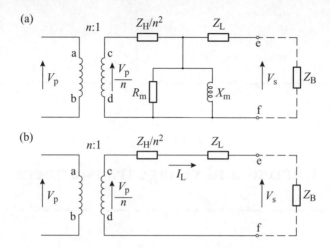

Figure 4.1   Voltage transformer equivalent circuits: (a) equivalent circuit; (b) simplified circuit

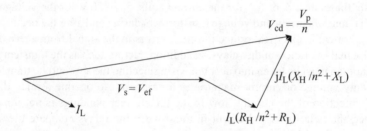

Figure 4.2   Vector diagram for a voltage transformer

branch can be ignored and the equivalent circuit then reduces to that shown in Figure 4.1b.

The vector diagram for a VT is given in Figure 4.2, with the length of the voltage drops increased for clarity. The secondary voltage V_s lags the voltage V_p/n and is smaller in magnitude. In spite of this, the nominal maximum errors are relatively small. VTs have an excellent transient behaviour and accurately reproduce abrupt changes in the primary voltage.

4.1.2 Errors

When used for measurement instruments, for example for billing and control purposes, the accuracy of a VT is important, especially for those values close to the nominal system voltage. Notwithstanding this, although the precision requirements of a VT for protection applications are not so high at nominal voltages, due to the problems of having to cope with a variety of different relays, secondary wiring burdens and the uncertainty of system parameters, errors should be contained within narrow limits over a wide range of possible voltages under fault conditions. This range should be between 5 and 173 per cent of the nominal primary voltage for VTs connected between line and earth.

Referring to the circuit in Figure 4.1a, errors in a VT are due to differences in magnitude and phase between V_p/n and V_s. These consist of the errors under open-circuit conditions when the load impedance Z_B is infinite, caused by the drop in voltage from the circulation of the magnetisation current through the primary winding, and errors due to voltage drops as a result of the load current I_L flowing through both windings. Errors in magnitude can be calculated from $Error_{VT} = \{(nV_s - V_p)/V_p\} \times 100\%$. If the error is positive, then the secondary voltage exceeds the nominal value.

4.1.3 Burden

The standard burden for voltage transformers is usually expressed in volt-amperes (VA) at a specified power factor.

Table 4.1 gives standard burdens based on ANSI Standard C57.13. Voltage transformers are specified in IEC publication 186A by the precision class, and the value of volt-amperes (VA).

The allowable error limits corresponding to different class values are shown in Table 4.2, where V_n is the nominal voltage. The phase error is considered positive when the secondary voltage leads the primary voltage. The voltage error is the percentage difference between the voltage at the secondary terminals, V_2, multiplied by the nominal transformation ratio, and the primary voltage V_1.

4.1.4 Selection of VTs

Voltage transformers are connected between phases, or between phase and earth. The connection between phase and earth is normally used with groups of three single-phase units connected in star at substations operating with voltages at about 34.5 kV or higher, or when it is necessary to measure the voltage and power factor of each phase separately.

The nominal primary voltage of a VT is generally chosen with the higher nominal insulation voltage (kV), and the nearest service voltage in mind. The nominal secondary voltages are generally standardised at 115 and 120 V. In order to select the nominal power of a VT, it is usual to add together all the nominal VA loadings of the apparatus connected to the VT secondary winding. In addition, it is important to take account of the voltage drops in the secondary wiring, especially if the distance between the transformers and the relays is large.

4.1.5 Capacitor voltage transformers

In general, the size of an inductive VT is proportional to its nominal voltage and, for this reason, the cost increases in a similar manner to that of a high voltage transformer. One alternative, and a more economic solution, is to use a capacitor voltage

Table 4.1   Standard burdens for voltage transformers

Standard burden			Characteristics for 120 V and 60 Hz			Characteristics for 69.3 V and 60 Hz		
design	volt-amperes	power factor	resistance (Ω)	inductance (H)	impedance (Ω)	resistance (Ω)	inductance (H)	impedance (Ω)
W	12.50	0.10	115.2	3.04	1152	38.4	1.01	384
X	25.00	0.70	403.2	1.09	575	134.4	0.364	192
Y	75.00	0.85	163.2	0.268	192	54.4	0.089	64
Z	200.00	0.85	61.2	0.101	72	20.4	0.034	24
ZZ	400.00	0.85	31.2	0.0403	36	10.2	0.0168	12
M	35.00	0.20	82.3	1.070	411	27.4	0.356	137

Table 4.2   Voltage transformer error limits

Class	Primary voltage	Voltage error (±%)	Phase error (±min)
0.1		0.1	0.5
0.2	$0.8V_n$, $1.0V_n$ and	0.2	10.0
0.5	$1.2V_n$	0.5	20.0
1.0		1.0	40.0
0.1		1.0	40.0
0.2	$0.05V_n$	1.0	40.0
0.5		1.0	40.0
1.0		2.0	80.0
0.1		0.2	80.0
0.2	V_n	2.0	80.0
0.5		2.0	80.0
1.0		3.0	120.0

transformer. This device is effectively a capacitance voltage divider and is similar to a resistive divider in that the output voltage at the point of connection is affected by the load – in fact the two parts of the divider taken together can be considered as the source impedance which produces a drop in voltage when the load is connected.

The capacitor divider differs from the inductive divider in that the equivalent impedance of the source is capacitive and the fact that this impedance can be compensated for by connecting a reactance in series at the point of connection. With an ideal reactance there are no regulation problems – however, in an actual situation on a network some resistance is always present. The divider can reduce the voltage to a value that enables errors to be kept within normally acceptable limits. For improved accuracy a high voltage capacitor is used in order to obtain a bigger voltage at the point of connection, which can be reduced to a standard voltage using a relatively inexpensive transformer as shown in Figure 4.3.

A simplified equivalent circuit of a capacitor VT is shown in Figure 4.4 in which V_i is equal to the nominal primary voltage, C is the numerically equivalent impedance equal to $(C_1 + C_2)$, L is the resonance inductance, R_i represents the resistance of the primary winding of transformer T plus the losses in C and L, and Z_e is the magnetisation impedance of transformer T. Referred to the intermediate voltage, the resistance of the secondary circuit and the load impedance are represented by R'_s and Z'_B, respectively, while V'_s and I'_s represent the secondary voltage and current.

It can be seen that, with the exception of C, the circuit in Figure 4.4 is the same as the equivalent circuit of a power transformer. Therefore, at the system frequency when C and L are resonating and cancelling out each other, under stable system conditions the capacitor VT acts like a conventional transformer. R_i and R'_s are not large and, in addition, I_e is small compared to I'_s so that the vector difference between

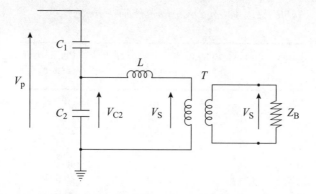

Figure 4.3   Capacitor VT basic circuit

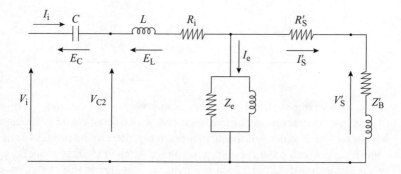

Figure 4.4   Capacitor VT equivalent circuit

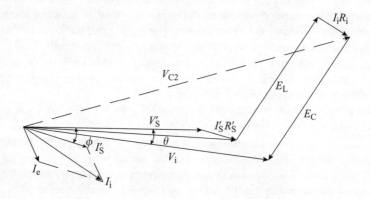

Figure 4.5   Capacitor VT vector diagram

V_i and V_s', which constitutes the error in the capacitor VT, is very small. This is illustrated in the vector diagram shown in Figure 4.5, which is drawn for a power factor close to unity. The voltage error is the difference in magnitude between V_i and V_s', whereas the phase error is indicated by the angle Θ. From the diagram it can be seen that, for frequencies different from the resonant frequency, the values of E_L and E_C predominate, causing serious errors in magnitude and phase.

Capacitor VTs have a better transient behaviour than electromagnetic VTs since the inductive and capacitive reactances in series are large in relation to the load impedance referred to the secondary voltage and thus, when the primary voltage collapses, the secondary voltage is maintained for some milliseconds because of the combination of the series and parallel resonant circuits represented by L, C and the transformer T.

4.2   Current transformers

Even though the performance required from a current transformer (CT) varies with the type of protection, high grade CTs must always be used. Good quality CTs are more reliable and result in fewer application problems and, in general, provide better protection. The quality of CTs is very important for differential protection schemes where the operation of the relays is directly related to the accuracy of the CTs under fault conditions as well as under normal load conditions.

CTs can become saturated at high current values caused by nearby faults; to avoid this, care should be taken to ensure that under the most critical faults the CT operates on the linear portion of the magnetisation curve. In all these cases the CT should be able to supply sufficient current so that the relay operates satisfactorily.

4.2.1   Equivalent circuit

An approximate equivalent circuit for a CT is given in Figure 4.6a, where $n^2 Z_H$ represents the primary impedance Z_H referred to the secondary side, and the secondary impedance is Z_L. R_m and X_m represent the losses and the excitation of the core.

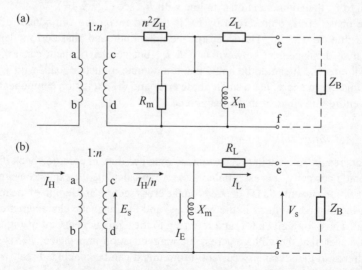

Figure 4.6   Current transformer equivalent circuits

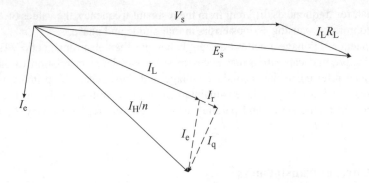

Figure 4.7   Vector diagram for the CT equivalent circuit

The circuit in Figure 4.6a can be reduced to the arrangement shown in Figure 4.6b where Z_H can be ignored, since it does not influence either the current I_H/n or the voltage across X_m. The current flowing through X_m is the excitation current I_e.

The vector diagram, with the voltage drops exaggerated for clarity, is shown in Figure 4.7. In general Z_L is resistive and I_e lags V_s by 90°, so that I_e is the principal source of error. Note that the net effect of I_e is to make I_L lag and be much smaller than I_H/n, the primary current referred to the secondary side.

4.2.2   Errors

The causes of errors in a CT are quite different to those associated with VTs. In effect, the primary impedance of a CT does not have the same influence on the accuracy of the equipment – it only adds an impedance in series with the line, which can be ignored. The errors are principally due to the current that circulates through the magnetising branch. The magnitude error is the difference in magnitude between I_H/n and I_L and is equal to I_r, the component of I_e in line with I_L (see Figure 4.7).

The phase error, represented by Θ, is related to I_q, the component of I_e that is in quadrature with I_L. The values of the magnitude and phase errors depend on the relative displacement between I_e and I_L, but neither of them can exceed the vectorial error I_e. It should be noted that a moderate inductive load, with I_e and I_L approximately in phase, has a small phase error and the excitation component results almost entirely in an error in the magnitude.

4.2.3   AC saturation

CT errors result from excitation current, so much so that, in order to check if a CT is functioning correctly, it is essential to measure or calculate the excitation curve. The magnetisation current of a CT depends on the cross section and length of the magnetic circuit, the number of turns in the windings, and the magnetic characteristics of the material. Thus, for a given CT, and referring to the equivalent circuit of Figure 4.6b, it can be seen that the voltage across the magnetisation impedance, E_s, is directly proportional to the secondary current. From this it can be concluded that, when the primary current and therefore the secondary current is increased, these currents reach a point when the core commences to saturate and the magnetisation current becomes sufficiently high enough to produce an excessive error.

When investigating the behaviour of a CT, the excitation current should be measured at various values of voltage – the so-called secondary injection test. Usually it is more convenient to apply a variable voltage to the secondary winding, leaving the primary winding open-circuited. Figure 4.8a shows the typical relationship between the secondary voltage and the excitation current, determined in this way. In European standards the point K_p on the curve is called the saturation or knee point and is defined as the point at which an increase in the excitation voltage of ten per cent produces an increase of 50 per cent in the excitation current. This point is referred to in the ANSI/IEEE standards as the intersection of the excitation curves with a 45° tangent line as indicated in Figure 4.8b. The European knee point is at a higher voltage than the ANSI/IEEE knee point.

4.2.4 Burden

The burden of a CT is the value in ohms of the impedance on the secondary side of the CT due to the relays and the connections between the CT and the relays. By way of example, the standard burdens for CTs with a nominal secondary current of 5 A are shown in Table 4.3, based on ANSI Standard C57.13.

IEC Standard Publication 185(1987) specifies CTs by the class of accuracy followed by the letter M or P, which denotes if the transformer is suitable for measurement or protection purposes respectively. The current and phase error limits for measurement and protection CTs are given in Tables 4.4a and 4.4b. The phase error is considered positive when the secondary current leads the primary current. The current error is the percentage deviation of the secondary current multiplied by the nominal transformation ratio, from the primary current, i.e., $\{(CTR \times I_2) - I_1\} \div I_1 (\%)$, where $I_1 = $ primary current (A), $I_2 = $ secondary current (A), and CTR = current transformer transformation ratio. Those CT classes marked with 'ext' denote wide range (extended) current transformers with a rated continuous current of 1.2 or two times the nameplate current rating.

4.2.5 Selection of CTs

When selecting a CT, it is important to ensure that the fault level and normal load conditions do not result in saturation of the core and that the errors do not exceed acceptable limits. These factors can be assessed from:

- formulae;
- CT magnetisation curves;
- CT classes of accuracy.

The first two methods provide precise facts for the selection of the CT. The third only provides a qualitative estimation. The secondary voltage E_s in Figure 4.6b has to be determined for all three methods. If the impedance of the magnetic circuit, X_m, is

(a)

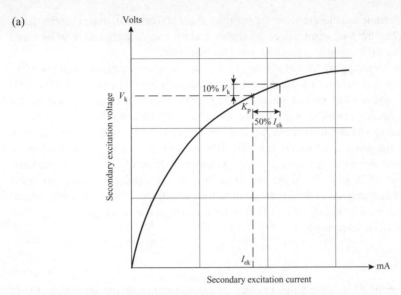

(b)

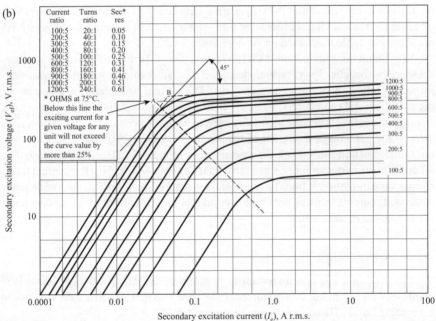

Figure 4.8   CT magnetisation curves: (a) defining the knee point in a CT excitation curve according to European standards; (b) typical excitation curves for a multi-ratio class C CT (from IEEE Standard C576.13-1978; reproduced by permission of the IEEE)

Table 4.3   Standard burdens for protection CTs with 5 A secondary current

Designation	Resistance (Ω)	Inductance (mH)	Impedance (Ω)	Volt-amps (at 5 A)	Power factor
B-1	0.5	2.3	1.0	25	0.5
B-2	1.0	4.6	2.0	50	0.5
B-4	2.0	9.2	4.0	100	0.5
B-8	4.0	18.4	8.0	200	0.5

Table 4.4a   Error limits for measurement current transformers

Class	% current error at the given proportion of rated current shown below							% phase error at the given proportion of the rated current shown below						
	2.0*	1.2	1.0	0.5	0.2	0.1	0.05	2.0*	1.2	1.0	0.5	0.2	0.1	0.05
0.1	–	0.1	0.1	–	0.20	0.25	–	–	5	5	–	8	10	–
0.2	–	0.2	0.2	–	0.35	0.50	–	–	10	10	–	15	20	–
0.5	–	0.5	0.5	–	0.75	1.00	–	–	30	30	–	45	60	–
1.0	–	1.0	1.0	–	1.50	2.00	–	–	60	60	–	90	120	–
3.0	–	3.0	–	3.0	–	–	–	–	120	–	120	–	–	–
0.1 ext	0.1	–	0.1	–	0.20	0.25	0.40	5	–	5	–	8	10	15
0.2 ext	0.2	–	0.2	–	0.35	0.50	0.75	10	–	10	–	15	20	30
0.5 ext	0.5	–	0.5	–	0.75	1.00	1.50	30	–	30	–	45	60	90
1.0 ext	1.0	–	1.0	–	1.50	2.00	–	60	–	60	–	90	120	–
3.0 ext	3.0	–	–	3.0	–	–	–	120	–	–	120	–	–	–

*ext = 200%

Table 4.4b   Error limits for protection current CTs

Class	Current error (%) at proportion of rated primary current shown				Phase error (minutes) at proportion of rated primary current shown			
	1.0	0.5	0.2	0.1	1.0	0.5	0.2	0.1
5P and 5P ext	1.0	–	1.5	2.0	60	–	90	120
10P and 10P ext	3.0	3.0	–	–	120	120	–	–

Total error for nominal error limit current and nominal load is 5% for 5P and 5P ext CTs and 10% for 10P and 10P ext CTs.

high, this can be removed from the equivalent circuit with little error, giving $E_s = V_s$, and thus:

$$V_s = I_L(Z_L + Z_C + Z_B) \tag{4.1}$$

where: V_s = r.m.s voltage induced in the secondary winding; I_L = maximum secondary current in amperes (this can be determined by dividing the maximum fault current on the system by the transformer turns ratio selected); Z_B = external impedance connected; Z_L = impedance of the secondary winding; Z_C = impedance of the connecting wiring.

Use of the formula

This method utilises the fundamental transformer equation:

$$V_s = 4.44 f A N B_{max} 10^{-8} \text{ volts} \tag{4.2}$$

where: f = frequency in Hz; A = cross-sectional area of core (in^2); N = number of turns; B_{max} = flux density (lines/in^2).

The cross-sectional area of metal and the saturation flux density are sometimes difficult to obtain. The latter can be taken as equal to 100 000 lines/in^2, which is a typical value for modern transformers. To use the formula, V_s is determined from eqn. 4.1, and B_{max} is then calculated using eqn. 4.2. If B_{max} exceeds the saturation density, there could be appreciable errors in the secondary current and the CT selected would not be appropriate.

Example 4.1

Assume that a CT with a ratio of 2000/5 is available, having a steel core of high permeability, a cross-sectional area of 3.25 in^2, and a secondary winding with a resistance of 0.31 Ω. The impedance of the relays, including connections, is 2 Ω. Determine whether the CT would be saturated by a fault of 35 000 A at 50 Hz.

Solution

If the CT is not saturated, then the secondary current, I_L, is $35\,000 \times 5/2000 = 87.5$ A. $N = 2000/5 = 400$ turns, and $V_s = 87.5 \times (0.31 + 2) = 202.1$ V. Using eqn. 4.2, B_{max} can now be calculated:

$$B_{max} = \frac{202.1 \times 10^8}{4.44 \times 50 \times 3.25 \times 400} = 70\,030 \text{ lines/in}^2$$

Since the transformer in this example has a steel core of high permeability this relatively low value of flux density should not result in saturation.

Using the magnetisation curve

Typical CT excitation curves, which are supplied by manufacturers, state the r.m.s. current obtained on applying an r.m.s. voltage to the secondary winding, with the primary winding open-circuited. The curves give the magnitude of the excitation current required in order to obtain a specific secondary voltage. The method consists

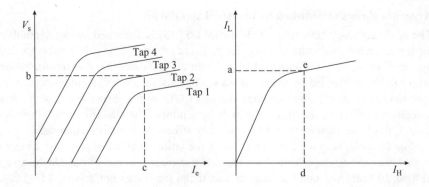

Figure 4.9   *Using the magnetisation curve:* a: assume a value for I_L; b: $V_s = I_L(Z_L + Z_c + Z_B)$; c: find I_e from the curve; d: $I_H = n(I_L + I_e)$; e: draw the point on the curve

of producing a curve that shows the relationship between the primary and secondary currents for one tap and specified load conditions, such as shown in Figure 4.9.

Starting with any value of secondary current, and with the help of the magnetisation curves, the value of the corresponding primary current can be determined. The process is summarised in the following steps:

(a)   Assume a value for I_L.

(b)   Calculate V_s in accordance with eqn. 4.1.

(c)   Locate the value of V_s on the curve for the tap selected, and find the associated value of the magnetisation current, I_e.

(d)   Calculate I_H/n $(= I_L + I_e)$ and multiply this value by n to refer it to the primary side of the CT.

(e)   This provides one point on the curve of I_L versus I_H. The process is then repeated to obtain other values of I_L and the resultant values of I_H. By joining the points together the curve of I_L against I_H is obtained.

This method incurs an error in calculating I_H/n by adding I_e and I_L together arithmetically and not vectorially, which implies not taking account of the load angle and the magnetisation branch of the equivalent circuit. However, this error is not great and the simplification makes it easier to carry out the calculations.

After constructing the curve it should be checked to confirm that the maximum primary fault current is within the transformer saturation zone. If not, then it will be necessary to repeat the process, changing the CT tap until the fault current is within the linear part of the characteristic. In practice it is not necessary to draw the complete curve because it is sufficient to take the known fault current and refer to the secondary winding, assuming that there is no saturation for the tap selected. This converted value can be taken as I_L initially for the process described earlier. If the tap is found to be suitable after finishing the calculations, then a value of I_H can be obtained that is closer to the fault current.

Accuracy classes established by the ANSI standards

The ANSI accuracy class of a CT (Standard C57.13) is described by two symbols – a letter and a nominal voltage; these define the capability of the CT. C indicates that the transformation ratio can be calculated, while T indicates that the transformation ratio can be determined by means of tests. The classification C includes those CTs with uniformly distributed windings and other CTs with a dispersion flux that has a negligible effect on the ratio, within defined limits. The classification T includes those CTs whose dispersion flux considerably affects the transformation ratio.

For example, with a CT of class C-100, the ratio can be calculated and the error should not exceed ten per cent if the secondary current does not go outside the range of 1 to 20 times the nominal current and if the load does not exceed $1\,\Omega\,(1\,\Omega\times 5\,\text{A}\times 20=100\,\text{V})$ at a minimum power factor of 0.5.

These accuracy classes are only applicable for complete windings. When considering a winding provided with taps, each tap will have a voltage capacity proportionally smaller, and in consequence it can only feed a portion of the load without exceeding the ten per cent error limit. The permissible load is defined as $Z_B = (N_P V_C)/100$, where Z_B is the permissible load for a given tap of the CT, N_P is the fraction of the total number of turns being used and V_C is the ANSI voltage capacity for the complete CT.

Example 4.2

The maximum fault current in a given circuit is 12 000 A. The nominal CT ratio is 1200/5 and the CT is to be used with a tap of 800/5. The CT class is C-200, the resistance of the secondary is $0.2\,\Omega$, the total secondary load is $2.4\,\Omega$ and the power factor is 0.6. Determine if, on the occurrence of a fault, the error will exceed ten per cent.

Solution

The resistance of the secondary winding of the CT can be ignored since, by definition, the class C-200 indicates that the CT could withstand 200 V plus the drop produced by the resistance of the secondary with a current range equal to 20 times the nominal value and with a load power factor as low as 0.5. Notwithstanding this, the voltage drops in the secondary can be ignored only if the current does not exceed 100 A. For the example given, $I_L = 12\,000 \times (5/800) = 75\,\text{A}$.

The permissible load is given by

$$Z_B = (N_P V_C) \div 100$$

$$N_P = 800/1200 = 0.667$$

so that

$$Z_B = (0.667 \times 200\,\text{V}) \div 100\,\text{A} = 1.334\,\Omega$$

Since the loading of the circuit, $2.4\,\Omega$, is more than the maximum permissible ($1.33\,\Omega$), then the error could exceed ten per cent during a fault of 12 000 A, which results in a maximum secondary current of 75 A. Consequently, it is necessary to reduce the load, increase the current transformer tap or use another CT of a higher class.

4.2.6 DC saturation

Up to now, the behaviour of a CT has been discussed in terms of a steady state, without considering the DC transient component of the fault current. However, the DC component has more influence in producing severe saturation than the AC component.

Figure 4.10 shows an example of the distortion and reduction in the secondary current that can be caused by DC saturation. However, the DC component of the fault current does not produce saturation of the CT if $V_K \geq 6.28\,IRT$, where: V_K = voltage at the knee point of the magnetisation curve, determined by the extension of the straight part of the curve; I = secondary symmetrical current (amperes – r.m.s.); R = total resistance of the secondary; T = DC time constant of the primary current in cycles, i.e.

$$T = \frac{L_p}{R_p} f$$

where: L_p = inductance of the primary circuit; R_p = resistance of the primary circuit; f = frequency.

DC saturation is particularly significant in complex protection schemes since, in the case of external faults, high fault currents circulate through the CTs. If saturation occurs in different CTs associated with a particular relay arrangement, this could result in the circulation of unbalanced secondary currents that would cause the system to malfunction.

4.2.7 Precautions when working with CTs

Working with CTs associated with energised network circuits can be extremely hazardous. In particular, opening the secondary circuit of a CT could result in dangerous

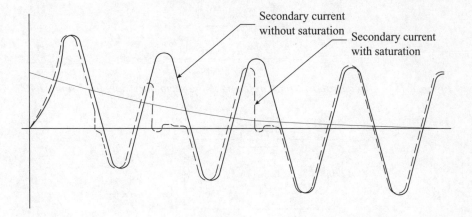

Figure 4.10   The effect of DC saturation on the secondary current

overvoltages, which might harm operational staff or lead to equipment being damaged because the current transformers are designed to be used in power circuits that have an impedance much greater than their own. As a consequence, when secondary circuits are left open, the equivalent primary-circuit impedance is almost unaffected but a high voltage will be developed by the primary current passing through the magnetising impedance. Thus, secondary circuits associated with CTs must always be kept in a closed condition or short-circuited in order to prevent these adverse situations occurring. To illustrate this, an example is given next using typical data for a CT and a 13.2 kV feeder.

Example 4.3

Consider a 13.2 kV feeder that is carrying a load of 10 MVA at 1.0 power factor. Associated with this circuit is a 500/5 CT feeding a measurement system whose total load is 10 VA. The equivalent circuit of the CT referred to the secondary side is shown in Figure 4.11. Calculate the voltage that would occur in the secondary circuit of the CT if the measurement system was accidentally opened.

Solution

The single line diagram is given in Figure 4.12 and the equivalent circuit in Figure 4.13.

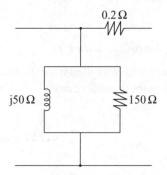

Figure 4.11   CT equivalent circuit, referred to the secondary side, for Example 4.3

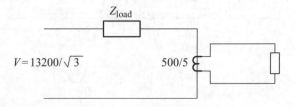

Figure 4.12   Single-line diagram for Example 4.3

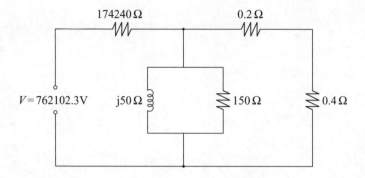

Figure 4.13   Equivalent circuit for Figure 4.12

Referring the values to the secondary side of the CT gives

$$V = \frac{13200}{\sqrt{3}} \times \frac{500}{5} = 762102.36\,\text{V}$$

$$Z_{\text{load}} = \frac{13.2^2}{10} \times \left(\frac{500}{5}\right)^2 = 174240\,\Omega$$

$$Z_{\text{meter}} = \frac{10}{5^2} = 0.4\,\Omega$$

When the secondary circuit is closed, the voltage across the measurement system can be calculated approximately, ignoring the shunt branch, as

$$V_{\text{meter}} = \frac{762102.36}{174240+0.2+0.4}\,\text{A} \times 0.4\,\Omega = 4.37\,\text{A} \times 0.4\,\Omega = 1.75\,\text{V}$$

If the secondary circuit is opened, the current is only able to circulate across the shunt branch. In these conditions the voltage that appears at the terminals of the CT is

$$V_{\text{CT}} = \frac{762102.36}{174240+(150\,\|\,j50)} \times (150\,\|\,j50) = 207.47\angle 71.55°\,\text{V}$$

Therefore the voltage increases by almost 120 times.

Chapter 5

Overcurrent protection

5.1 General

Very high current levels in electrical power systems are usually caused by faults on the system. These currents can be used to determine the presence of faults and operate protection devices, which can vary in design depending on the complexity and accuracy required. Among the more common types of protection are thermo-magnetic switches, moulded-case circuit breakers (MCCBs), fuses, and overcurrent relays. The first two types have simple operating arrangements and are principally used in the protection of low voltage equipment. Fuses are also often used at low voltages, especially for protecting lines and distribution transformers.

Overcurrent relays, which form the basis of this chapter, are the most common form of protection used to deal with excessive currents on power systems. They should not be installed purely as a means of protecting systems against overloads – which are associated with the thermal capacity of machines or lines – since overcurrent protection is primarily intended to operate only under fault conditions. However, the relay settings that are selected are often a compromise in order to cope with both overload and overcurrent conditions.

5.2 Types of overcurrent relay

Based on the relay operating characteristics, overcurrent relays can be classified into three groups: definite current or instantaneous, definite time, and inverse time. The characteristic curves of these three types are shown in Figure 5.1, which also illustrates the combination of an instantaneous relay with one having an inverse time characteristic.

5.2.1 Definite-current relays

This type of relay operates instantaneously when the current reaches a predetermined value. The setting is chosen so that, at the substation furthest away from the source,

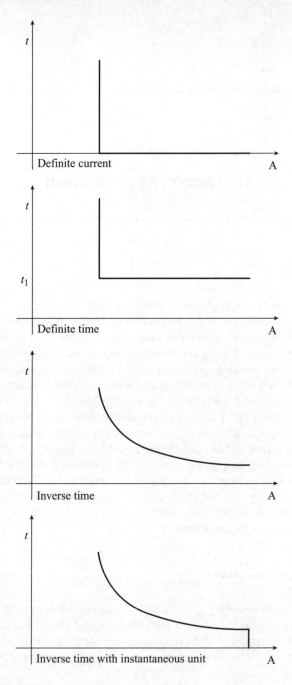

Figure 5.1   Time/current operating characteristics of overcurrent relays

the relay will operate for a low current value and the relay operating currents are progressively increased at each substation, moving towards the source. Thus, the relay with the lower setting operates first and disconnects load at the point nearest to the fault. This type of protection has the drawback of having little selectivity at high values of short-circuit current. Another disadvantage is the difficulty of distinguishing between the fault current at one point or another when the impedance between these points is small in comparison to the impedance back to the source, leading to the possibility of poor discrimination.

Figure 5.2a illustrates the effect of the source impedance on the short-circuit level at a substation, and for a fault at point B down the line. From Figure 5.2b it can be appreciated that the fault currents at F_1 and F_2 are almost the same, and it is this that makes it difficult to obtain correct settings for the relays. When there is some

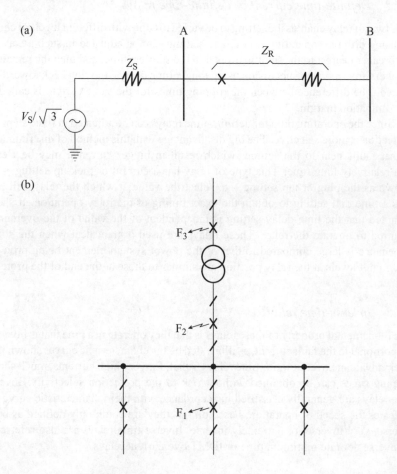

Figure 5.2   *Illustration of different levels of fault current.* (a) $Z_R =$ impedance of protected element. (b) $Z_S =$ source impedance. $I_{sc(A)} = (V_S/\sqrt{3}) \times (1/Z_S)$, $I_{sc(B)} = V_S/(\sqrt{3}(Z_S + Z_R))$

considerable impedance between F_1 and F_2, for example when the fault F_1 is located down a long line, then the fault current at F_1 will be less than at F_2. Similarly, due to the impedance of the transformer, there will be a considerable difference between the currents for faults at F_2 and F_3, even though these two points are physically close.

If the protection settings are based on maximum fault level conditions, then these settings may not be appropriate for the situation when the fault level is lower. However, if a lower value of fault level is used when calculating the relay settings, this could result in some breakers operating unnecessarily if the fault level increases. As a consequence, definite current relays are not used as the only overcurrent protection, but their use as an instantaneous unit is common where other types of protection are in use.

5.2.2   Definite-time/current or definite-time relays

This type of relay enables the setting to be varied to cope with different levels of current by using different operating times. The settings can be adjusted in such a way that the breaker nearest to the fault is tripped in the shortest time, and then the remaining breakers are tripped in succession using longer time delays, moving back towards the source. The difference between the tripping times for the same current is called the discrimination margin.

Since the operating time for definite-time relays can be adjusted in fixed steps, the protection is more selective. The big disadvantage with this method of discrimination is that faults near to the source, which result in bigger currents, may be cleared in a relatively long time. This type of relay has a current or pick-up setting – also known as the plug or tap setting – to select the value at which the relay will start, plus a time dial setting to obtain the exact timing of the relay operation. It should be noted that the time-delay setting is independent of the value of the overcurrent required to operate the relay. These relays are used a great deal when the source impedance is large compared to that of the power system element being protected when fault levels at the relay position are similar to those at the end of the protected element.

5.2.3   Inverse-time relays

The fundamental property of these relays is that they operate in a time that is inversely proportional to the fault current, as illustrated by the characteristic curves shown later. Their advantage over definite-time relays is that, for very high currents, much shorter tripping times can be obtained without risk to the protection selectivity. Inverse-time relays are generally classified in accordance with their characteristic curve that indicates the speed of operation; based on this they are commonly defined as being inverse, very inverse, or extremely inverse. Inverse-time relays are also referred to as inverse definite minimum time or IDMT overcurrent relays.

5.3   Setting overcurrent relays

Overcurrent relays are normally supplied with an instantaneous element and a time-delay element within the same unit. When electromechanical relays were more popular, the overcurrent protection was made up from separate single-phase units. The more modern microprocessor protection has a three-phase overcurrent unit and an earth-fault unit within the same case. Setting overcurrent relays involves selecting the parameters that define the required time/current characteristic of both the time-delay and instantaneous units. This process has to be carried out twice, once for the phase relays and then repeated for the earth-fault relays. Although the two processes are similar, the three-phase short-circuit current should be used for setting the phase relays while the phase-to-earth fault current should be used for the earth-fault relays. When calculating the fault currents, the power system is assumed to be in its normal operating state. However, at a busbar that has two or more transformers connected in parallel and protected with relays that do not have the facility of multiple setting groups – the ability to be adjusted to accommodate the prevailing system conditions, which is possible with numerical relays for example – then better discrimination is obtained if the calculations are carried out on the basis of each one of the transformers being out of service in turn. The same procedure can be applied to multiple circuit arrangements.

5.3.1 Setting instantaneous units

Instantaneous units are more effective when the impedances of the power system elements being protected are large in comparison to the source impedance, as indicated earlier. They offer two fundamental advantages:

- they reduce the operating time of the relays for severe system faults;
- they avoid the loss of selectivity in a protection system consisting of relays with different characteristics; this is obtained by setting the instantaneous units so that they operate before the relay characteristics cross, as shown in Figure 5.3.

The criteria for setting instantaneous units vary depending on the location, and the type of system element being protected. Three groups of elements can be defined: lines between substations, distribution lines, and transformers.

(i) Lines between substations

The setting of instantaneous units is carried out by taking at least 125 per cent of the symmetrical r.m.s. current for the maximum fault level at the next substation. The

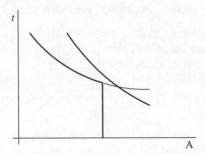

Figure 5.3   Preservation of selectivity using instantaneous units

procedure must be started from the furthest substation, then continued by moving back towards the source. When the characteristics of two relays cross at a particular system fault level, thus making it difficult to obtain correct co-ordination, it is necessary to set the instantaneous unit of the relay at the substation that is furthest away from the source to such a value that the relay operates for a slightly lower level of current, thus avoiding loss of co-ordination. The 25 per cent margin avoids overlapping the downstream instantaneous unit if a considerable DC component is present. In HV systems operating at 220 kV or above, a higher value should be used since the X/R ratio becomes larger, as does the DC component.

(ii) Distribution lines

The setting of the instantaneous element of relays on distribution lines that supply only pole-mounted MV/LV transformers is dealt with differently to the previous case, since these lines are at the end of the MV system. They therefore do not have to fulfil the co-ordination conditions that have to be met by lines between substations and so one of the following two values can be used to set these units:

1. 50 per cent of the maximum short-circuit current at the point of connection of the CT supplying the relay.
2. Between six and ten times the maximum circuit rating.

(iii) Transformer units

The instantaneous units of the overcurrent relays installed on the primary side of the transformers should be set at a value between 125 and 150 per cent of the short-circuit current existing at the busbar on the low voltage side, referred to the high voltage side. This value is higher that those mentioned previously to avoid lack of co-ordination with the higher currents encountered due to the magnetic inrush current when energising the transformer. If the instantaneous units of the transformer secondary winding overcurrent protection and the feeder relays are subjected to the same short-circuit level, then the transformer instantaneous units need to be overridden to avoid loss of selectivity unless there are communication links between these units that can permit the disabling of the transformer instantaneous overcurrent protection for faults detected by the feeder instantaneous overcurrent protection.

5.3.2   Coverage of instantaneous units protecting lines between substations

The percentage of coverage of an instantaneous unit which protects a line, X, can be illustrated by considering the system shown in Figure 5.4.

The following parameters are defined:

$$K_i = \frac{I_{pickup}}{I_{end}}$$

and

$$K_s = \frac{Z_{source}}{Z_{element}}$$

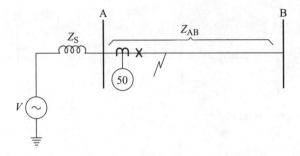

Figure 5.4   Coverage of instantaneous units

From Figure 5.4:

$$I_{pickup} = \frac{V}{Z_s + X Z_{ab}} \tag{5.1}$$

where: V = voltage at the relay CT point; Z_s = source impedance; Z_{ab} = impedance of the element being protected = $Z_{element}$; X = percentage of line protected; I_{end} = current at the end of the line; and $I_{pick\ up}$ = minimum current value for relay pick up.

$$I_{end} = \frac{V}{Z_s + Z_{ab}} \tag{5.2}$$

$$K_i = \frac{Z_s + Z_{ab}}{Z_s + X Z_{ab}} \Rightarrow X = \frac{Z_s + Z_{ab} - Z_s K_i}{Z_{ab} K_i} \tag{5.3}$$

This gives:

$$K_s = \frac{Z_s}{Z_{ab}} \Rightarrow X = \frac{K_s(1 - K_i) + 1}{K_i} \tag{5.4}$$

For example, if $K_i = 1.25$, and $K_s = 1$, then $X = 0.6$, i.e. the protection covers 60 per cent of the line.

Example 5.1

The effect of reducing the source impedance, Z_s, on the coverage provided by the instantaneous protection can be appreciated by considering the system in Figure 5.5, and using a value of 1.25 for K_i in eqn. 5.4. From this:

Z_S (Ω)	Z_{AB} (Ω)	I_A (A)	I_B (A)	% coverage
10	10	100	50	60
2	10	500	83	76

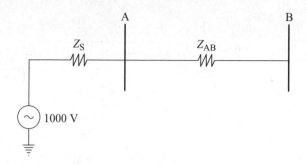

Figure 5.5   Equivalent circuit for Example 5.1

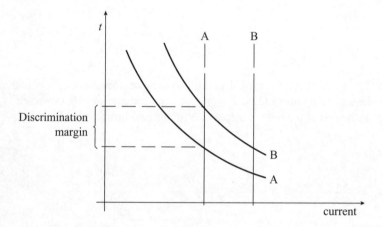

Figure 5.6   Overcurrent inverse-time relay curves associated with two breakers on the same feeder

5.3.3   Setting the parameters of time delay overcurrent relays

The operating time of an overcurrent relay has to be delayed to ensure that, in the presence of a fault, the relay does not trip before any other protection situated closer to the fault. The curves of inverse-time overcurrent relays associated with two breakers on the same feeder in a typical system are shown in Figure 5.6, illustrating the difference in the operating time of these relays at the same fault levels in order to satisfy the discrimination margin requirements. Definite-time relays and inverse-time relays can be adjusted by selecting two parameters – the time dial or time multiplier setting, and the pick-up or plug setting (tap setting).

The pick-up setting

The pick-up setting, or plug setting, is used to define the pick-up current of the relay, and fault currents seen by the relay are expressed as multiples of this. This value is usually referred to as the plug setting multiplier (PSM), which is defined as the ratio of the fault current in secondary amps to the relay pick-up or plug setting. For phase relays the pick-up setting is determined by allowing a margin for overload above the nominal current, as in the following expression:

$$\text{Pick-up setting} = (\text{OLF} \times I_{\text{nom}}) \div \text{CTR} \qquad (5.5)$$

where: OLF = overload factor that depends on the element being protected; I_{nom} = nominal circuit current rating; CTR = CT ratio.

The overload factor recommended for motors is 1.05. For lines, transformers and generators it is normally in the range of 1.25 to 1.5. In distribution systems where it is possible to increase the loading on feeders under emergency conditions, the overload factor can be of the order of 2. In any case I_{nom} has to be smaller than those of the CT and the thermal capacity of the conductor; otherwise the smallest value has to be taken to calculate the pick-up setting.

For earth-fault relays, the pick-up setting is determined taking account of the maximum unbalance that would exist in the system under normal operating conditions. A typical unbalance allowance is 20 per cent so that the expression in eqn. 5.5 becomes

$$\text{Pick-up setting} = (0.2 \times I_{\text{nom}}) \div \text{CTR} \qquad (5.6)$$

In HV transmission lines the unbalance allowance could go down to 10 per cent, while in rural distribution feeders the value could be as high as 30 per cent.

Time dial setting

The time dial setting adjusts the time delay before the relay operates whenever the fault current reaches a value equal to, or greater than, the relay current setting. In electromechanical relays the time delay is usually achieved by adjusting the physical distance between the moving and fixed contacts; a smaller time dial value results in shorter operating times. The time dial setting is also referred to as the time multiplier setting.

The criteria and procedures for calculating the time dial setting, to obtain the appropriate protection and co-ordination for the system, are considered next. These criteria are mainly applicable to inverse-time relays, although the same methodology is valid for definite-time relays.

1. Determine the required operating time t_1 of the relay furthest away from the source by using the lowest time dial setting and considering the fault level for which the instantaneous unit of this relay picks up. This time dial setting may have to be higher if the load that flows when the circuit is re-energised after a loss of supply is high (the cold load pick-up), or if it is necessary to co-ordinate with devices installed downstream, e.g. fuses or reclosers.

2. Determine the operating time of the relay associated with the breaker in the next substation towards the source, $t_{2a} = t_1 + t_{\text{margin}}$, where t_{2a} is the operating time of the back-up relay associated with breaker 2 and t_{margin} is the discrimination margin. The fault level used for this calculation is the same as that used to determine the timing t_1 of the relay associated with the previous breaker.

3. With the same fault current as in 1 and 2 above, and knowing t_{2a} and the pick-up value for relay 2, calculate the time dial setting for relay 2. Use the closest available relay time dial setting whose characteristic is above the calculated value.
4. Determine the operating time (t_{2b}) of relay 2, but now using the fault level just before the operation of its instantaneous unit.
5. Continue with the sequence, starting from the second stage.

The procedure referred to above is appropriate if it can be assumed that the relays have their characteristic curves scaled in seconds. For those relays where the time adjustment is given as a percentage of the operating curve for one second, the time dial setting can be determined starting from the fastest multiplier applied to the curve for time dial 1. In most modern relays the time settings can start from values as low as 0.1 s, in steps of 0.1 s.

Time discrimination margin

A time discrimination margin between two successive time/current characteristics of the order of 0.25 to 0.4 s should be typically used. This value avoids losing selectivity due to one or more of the following:

- breaker opening time;
- relay overrun time after the fault has been cleared;
- variations in fault levels, deviations from the characteristic curves of the relays (for example, due to manufacturing tolerances), and errors in the current transformers.

In numerical relays there is no overrun, and therefore the margin could be chosen as low as 0.2 s.

Single-phase faults on the star side of a Dy transformer are not seen on the delta side. Therefore, when setting earth-fault relays, the lowest available time dial setting can be applied to the relays on the delta side, which makes it possible to considerably reduce the settings and thus the operating times of earth-fault relays nearer the source infeed.

Use of mathematical expressions for the relay characteristics

The procedure indicated above for phase and earth units can easily be used when the operating characteristics of the relays are defined by mathematical formulae instead of by curves on log-log paper. IEC and ANSI/IEEE Standards define the operating time mathematically by the following expression:

$$t = \frac{k\beta}{(I/I_s)^\alpha - 1} + L \qquad (5.7)$$

where: t = relay operating time in seconds; k = time dial, or time multiplier, setting; I = fault current level in secondary amps; I_s = pick-up current selected; L = constant.

The constants α and β determine the slope of the relay characteristics. The values of α, β and L for various standard overcurrent relay types manufactured under ANSI/IEEE and IEC Standards are given in Table 5.1. Typical characteristics for both types are shown in Figures 5.7 and 5.8.

Table 5.1   *ANSI/IEEE and IEC constants for standard overcurrent relays*

Curve description	Standard	α	β	L
Moderately inverse	IEEE	0.02	0.0515	0.114
Very inverse	IEEE	2.0	19.61	0.491
Extremely inverse	IEEE	2.0	28.2	0.1217
Inverse	CO8	2.0	5.95	0.18
Short-time inverse	CO2	0.02	0.0239	0.0169
Standard inverse	IEC	0.02	0.14	0
Very inverse	IEC	1.0	13.5	0
Extremely inverse	IEC	2.0	80.0	0
Long-time inverse	UK	1.0	120	0

Given the relay characteristic, it is a straightforward task to calculate the time response for a given time dial setting k, pick-up setting, and the other values of the expression in eqn. 5.7. Likewise, if a particular time response and pick-up setting have been determined, the time dial setting is found by solving k from the same equation.

5.4   Constraints of relay co-ordination

5.4.1   Minimum short-circuit levels

When the time delay unit has been set, using maximum fault levels, it is necessary to check that the relays will operate at the minimum fault levels, and in the correct sequence. For this it is sufficient to verify that the plug setting multiplier – (I/I_s) in eqn. 5.7 – under these conditions is greater than 1.5.

5.4.2   Thermal limits

Once the curves for the overcurrent relays have been defined, a check should be made to ensure that they lie below the curves for the designated thermal capacity of machines and cables. In the case of conductors, manufacturers' graphs, which indicate the length of time that different sizes can withstand various short-circuit values, should be used. A typical graph for copper conductors with thermoplastic insulation is given in Figure 5.9. For motors, the manufacturers' information should also be consulted.

In the case of transformers, the magnitude of the fault current that they can withstand during a given time is limited by their impedance. ANSI/IEEE Standard 242-1986 defines curves of short-circuit capacity for four categories of liquid-immersed transformers, based on the nominal kVA rating of the transformer and the short-circuit impedance.

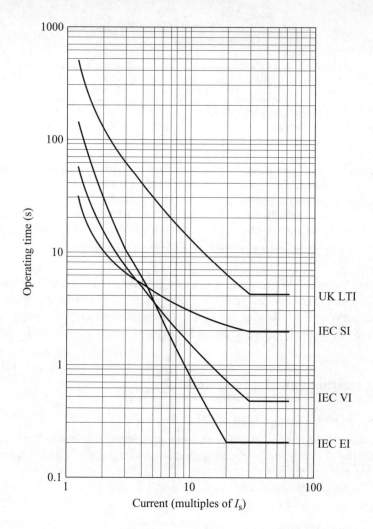

Figure 5.7   IEC overcurrent relay curves

Figures 5.10 to 5.13 show the curves of thermal capacity of transformers with the following characteristics:

 (i)   Category I – power rating between 5 and 500 kVA single phase; 15 to 500 kVA three phase.
 (ii) Category II – power rating between 501 and 1667 kVA single phase; 501 to 5000 kVA three phase.
(iii) Category III – power rating between 1668 and 10 000 kVA single phase; 5001 kVA to 30 000 kVA three phase.
(iv) Category IV – power rating above 10 000 kVA single phase; above 30 000 kVA three phase.

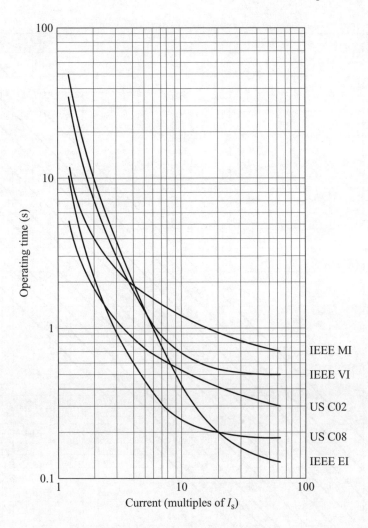

Figure 5.8   ANSI/IEEE overcurrent relay curves

The thermal limit curves for Dy transformers have to be shifted to the left by a ratio of $1/\sqrt{3}$ to make them more sensitive. This compensates for the lower value of current seen by the relays installed on the primary side, relative to the currents seen by the relays on the secondary side, during single-phase fault conditions, as discussed in Section 5.5.

5.4.3   Pick-up values

It is also important to check that the relay settings are not going to present problems when other system elements are energised. This is particularly critical for motors,

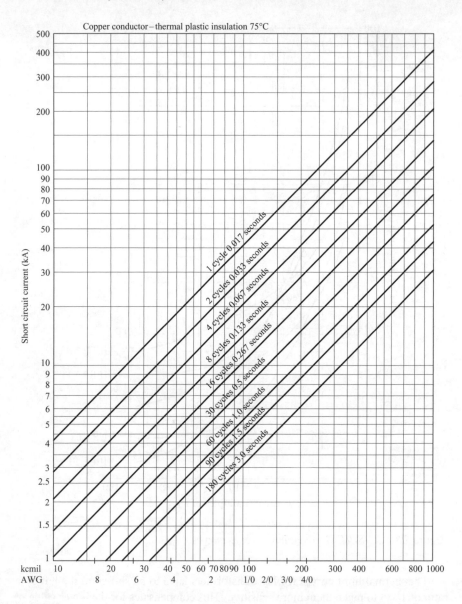

Figure 5.9   Thermal limits of copper conductors with thermoplastic insulation

and the appropriate code letter, which indicates the number of times nominal current taken when the motor is starting, should always be borne in mind.

In the case of transformers, the initial magnetisation inrush current that a transformer takes can be expressed as $I_{\text{Inrush}} = K \times I_{\text{nom}}$, where I_{nom} is the nominal transformer current, and the constant K depends on the transformer capacity; from

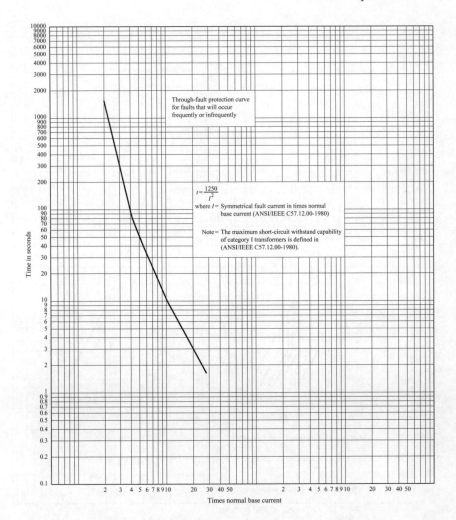

*Figure 5.10   Thermal capacity of transformers between 5 and 500 kVA single phase;
15 to 500 kVA three phase (from ANSI/IEEE Standard 242-1986;
reproduced by permission of the IEEE)*

500 to 2500 kVA, $K = 8$, and above 2500 kVA, $K = 10$. The inrush point then remains defined by the appropriate inrush current during 0.1 s.

Example 5.2

For the system shown in Figure 5.14, and starting from the data that are given there, carry out the following:

1. Calculate the nominal currents and three-phase short-circuit levels at each breaker.

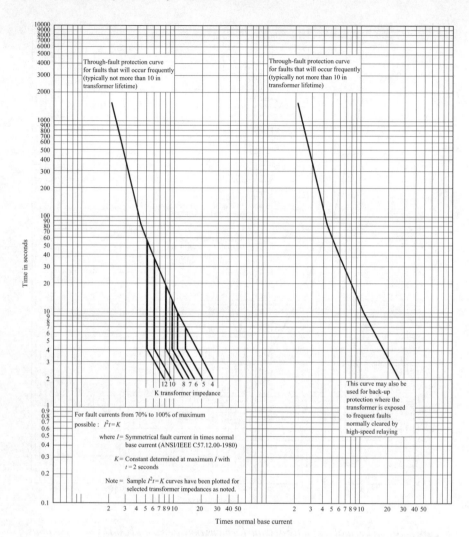

*Figure 5.11   Thermal capacity of transformers between 501 and 1667 single phase;
501 to 5000 kVA three phase (from ANSI/IEEE Standard 242-1968;
reproduced by permission of the IEEE)*

2. Select the transformation ratios of the CTs.
3. Determine the values of the pick-up setting, time dial and instantaneous settings
    of all phase relays to ensure a co-ordinated protection arrangement.
4. Find the percentage of the line BC protected by the instantaneous unit of the
    overcurrent relay associated with breaker 2.
5. Draw the time/current characteristics of the relays on the system.

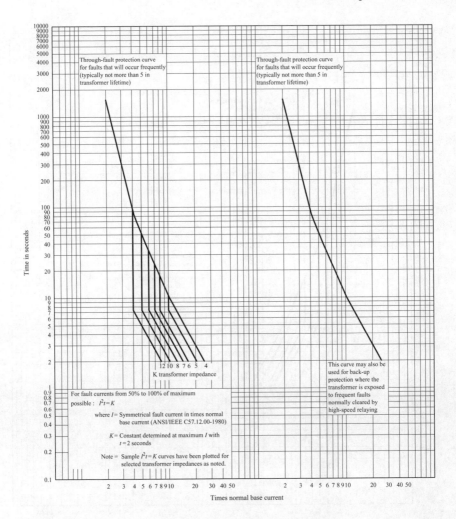

Figure 5.12   Thermal capacity of transformers between 1668 and 10000 kVA single phase; 5001 to 30000 kVA three phase (from ANSI/IEEE Standard 242-1968; reproduced by permission of the IEEE)

Take into account the following considerations:

1. The discrimination margin to be 0.4 s.
2. All relays have inverse time characteristics, as shown in Figure 5.15.
3. Relay data:
    Pick-up setting: 1 to 12 A in steps of 1 A
    Time dial setting: as in Figure 5.15
    Instantaneous: 6 to 144 A in steps of 1 A.

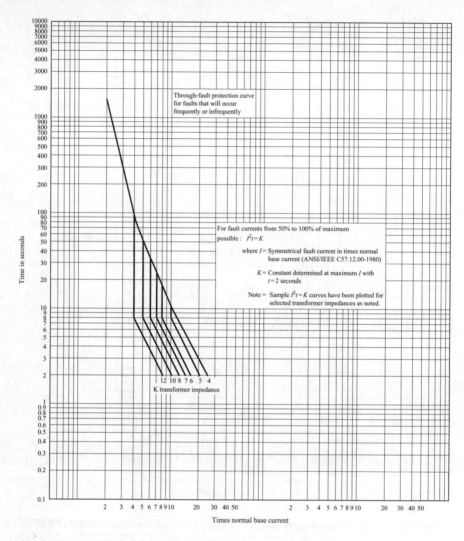

Figure 5.13 *Thermal capacity of transformers above 10000 kVA single phase;*
above 30000 kVA three phase (from ANSI/IEEE Standard 242-1968;
reproduced by permission of the IEEE)

Solution

Calculation of nominal currents and three-phase short-circuit levels

From Figure 5.14 the short-circuit level at busbar A, and the impedance of the line BC, can be obtained:

$$Z_{source} = \frac{V^2}{P_{sc}} = \frac{\left(115 + 10^3\right)^2}{950 \times 10^6} = 13.92 \,\Omega \text{ referred to } 115\,\text{kV}$$

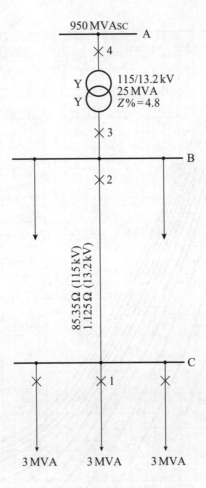

Figure 5.14   Schematic diagram for Example 5.2

$$Z_{\text{transf}} = Z_{\text{pu}} \times Z_{\text{Base}} = 0.048 \times \frac{\left(115 \times 10^3\right)^2}{25 \times 10^6} = 25.39\,\Omega \text{ referred to } 115\,\text{kV}$$

$$Z_{\text{lineBC}} = 85.35\,\Omega \text{ referred to } 115\,\text{kV}$$

The equivalent circuit of the system referred to 115 kV is shown in Figure 5.16.

Nominal currents

$$I_{\text{nom1}} = \frac{P}{\sqrt{3} \times V} = \frac{3 \times 10^6}{\sqrt{3}\left(13.2 \times 10^3\right)} = 131.2\,\text{A}$$

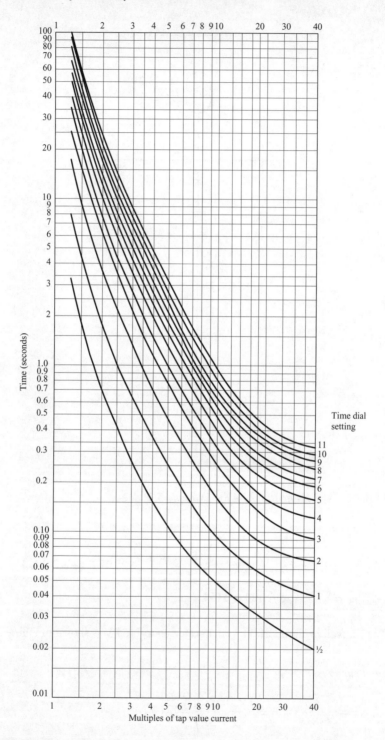

Figure 5.15   Typical operating curves for an inverse-time relay

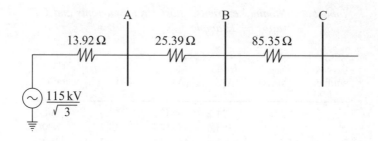

Figure 5.16   Equivalent circuit of the system shown in Figure 5.14

$$I_{nom2} = 3 \times I_{nom1} = 3 \times 131.2 = 393.6 \, A$$

$$I_{nom3} = \frac{25 \times 10^6}{\sqrt{3}\,(13.2 \times 10^3)} = 1093.5 \, A$$

$$I_{nom4} = \frac{25 \times 10^6}{\sqrt{3}\,(115 \times 10^3)} = I_{nom3} \times (13.2/115) = 125.5 \, A$$

Short-circuit levels

The equivalent circuit gives:

$$I_{faultC} = \frac{115 \times 10^3}{\sqrt{3}\,(13.92 + 25.39 + 83.35)} = 532.6 \, A \text{ referred to } 115 \, kV$$

$$= 532.6(115/13.2) = 4640.2 \, A \text{ referred to } 13.2 \, kV$$

$$I_{faultB} = \frac{115 \times 10^3}{\sqrt{3}\,(13.92 + 25.39)} = 1689.0 \, A \text{ referred to } 115 \, kV$$

$$= 1689(115/13.2) = 14714.8 \, A \text{ referred to } 13.2 \, kV$$

$$I_{faultA} = \frac{115 \times 10^3}{\sqrt{3} \times 13.92} = 4769.8 \, A \text{ referred to } 115 \, kV$$

Choice of CT transformer ratio

The transformation ratio of the CTs is determined by the larger of the two following values:

(i) I_{nom}
(ii) maximum short-circuit current without saturation being present. To fulfil this condition and assuming that a C100 core is used and that the total burden is 1 Ω, then $I_{sc}(5/X) \leq 100 \, A$ where I_{sc} is the short-circuit current.

Table 5.2 summarises the calculations.

Table 5.2   Nominal currents, short-circuit currents and CT ratios for Example 5.2

Breaker number	P_{nom} (MVA)	I_{nom} (A)	I_{sc} (A)	$(5/100)I_{sc}$ (A)	CT ratio
1	3	131.2	4640.0	232.0	300/5
2	9	393.6	14714.8	1735.7	800/5
3	25	1093.5	14714.8	735.7	1100/5
4	25	125.5	4769.8	238.5	300/5

Determination of the pick-up setting, time dial and instantaneous setting values: calculation of the pick-up settings

Relay 1: $1.5(131.2)5/300 = 3.28$ A; $\Rightarrow$ pick-up set at 4 A
Relay 2: $1.5(393.6)5/800 = 3.69$ A; $\Rightarrow$ pick-up set at 4 A
Relay 3: $1.5(1093.5)5/1100 = 7.46$ A; $\Rightarrow$ pick-up set at 8 A
Relay 4: $1.5(125.5)5/300 = 3.14$A; $\Rightarrow$ pick-up set at 4 A

Determination of time dial setting and calibration of the instantaneous setting

Relay 1
$I_{pick\ up} = 4 \times 300/5 = 240$ A
Time dial setting selected is 1.0
Setting of instantaneous element $= (0.5\ I_{sc})(1/CTR) = (0.5 \times 4640)\ 5/300 = 38.67$ A; set at 39 A.
$I_{inst.\ trip} = 39 \times 300/5 = 2340$ A primary at 13.2 kV.
Plug setting multiplier, $PSM_b = (2340\ A \times 5/300) \times 1/4 = 9.75$ times
From Figure 5.15, with a plug setting multiplier of 9.75 and a time dial setting at 1, $t_{1b} = 0.1$ s

Relay 2
2340 A should produce operation of t_{2a} in at least $0.1 + 0.4 = 0.5$ s.
$PSM_a = 2340\ A \times 5/800 \times 1/4 = 3.66$ times
With PSM at 3.66 times, and t_{op} at least 0.5 s, time dial $= 2$ is chosen
Instantaneous setting $= (1.25\ I_{faultC})(1/CTR) = 1.25\ (4640)(5/800) = 36.25$ A; set at 37 A
$I_{inst.prim} = (37)\ 800/5 = 5920$ A at 13.2 kV
$PSM_b = 5920\ A \times 5/800 \times 1/4 = 9.25$ times
With 9.25 PSM and time dial setting $= 2 \Rightarrow t_{2b} = 0.18$ s

Relay 3
To discriminate with relay 2, take $I_{inst.prim2} = 5920$ A
Require operation of t_{3a} in at least $0.18 + 0.4 = 0.58$ s
$PSM_a = 5920\ A \times 5/1100 \times 1/8 = 3.36$ times
With 3.36 PSM and $t_{op} = 0.58$ s, $\Rightarrow$ time dial setting $= 2$

However, the instantaneous element of the relay associated with breaker 3 is over-ridden and the discrimination time is applied for a fault on busbar B to avoid lack of co-ordination with the instantaneous units of the relays associated with the feeders from the busbar, as referred to in Section 5.3.1.

Based on $I_{sc} = 14714.8$ at 13.2 kV, $PSM_b = 14714.8$ A $\times 5/1100 \times 1/8 = 8.36$ times
With 8.36 times PSM and time dial setting $= 2, \Rightarrow t_{3b} = 0.21$ s

Relay 4
For 14714.8 A, PSM $= 14714.8$ A $(13.2/115)$ $5/300 \times 1/4 = 7.04$ times.
Require $t_4 = 0.21 + 0.4 = 0.61$ s
With 7.04 PSM and $t_{op} = 0.61$ s $\Rightarrow$ time dial setting $= 5$.

Setting of instantaneous element $= (1.25 \times I_{faultB})(1/CTR)$

$$= 1.25\,(1689)\,5/300$$

$$= 35.19\,\text{A}$$

Setting $= 36$ A.
$I_{inst.prim} = 36\,(300/5) = 2160$ A referred to 115 kV
$I_{inst.prim} = 2160\,(115/13.2) = 18818.2$ A referred to 13.2 kV

Table 5.3 summarises the four relay settings.

Percentage of line A-B protected by the instantaneous element of the relay associated with breaker 2

$$X\% = \frac{K_s\,(1 - K_i) + 1}{K_i}$$

$$K_i = \frac{I_{sc\ pick\ up}}{I_{sc\ end}} = \frac{5920}{4640} = 1.28$$

$$K_s = \frac{Z_{source}}{Z_{element}} = \frac{13.92 + 25.39}{85.35} = 0.46$$

and

$$X\% = \frac{0.46\,(1 - 1.28) + 1}{1.28} = 0.68$$

Therefore, the instantaneous element covers 68 per cent of the line BC.

Table 5.3   Summary of settings for Example 5.2

Relay associated with breaker number	Pick-up (A)	Time dial	Instantaneous I_{sec} (A)	Instantaneous I_{prim} (A)
1	2.5	–	40	2400
2	4.0	2	37	5920
3	8.0	2	–	–
4	4.0	5	36	18818

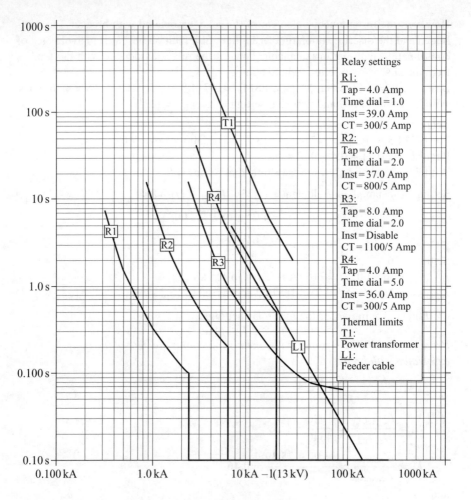

Relay settings

R1:
Tap = 4.0 Amp
Time dial = 1.0
Inst = 39.0 Amp
CT = 300/5 Amp

R2:
Tap = 4.0 Amp
Time dial = 2.0
Inst = 37.0 Amp
CT = 800/5 Amp

R3:
Tap = 8.0 Amp
Time dial = 2.0
Inst = Disable
CT = 1100/5 Amp

R4:
Tap = 4.0 Amp
Time dial = 5.0
Inst = 36.0 Amp
CT = 300/5 Amp

Thermal limits
T1:
Power transformer
L1:
Feeder cable

Figure 5.17   Relay co-ordination curves for Example 5.2

The co-ordination curves of the relays associated with this system are shown in Figure 5.17. It should be noted that these are all drawn for currents at the same voltage – in this case 13.2 kV.

5.5   Co-ordination across Dy transformers

In the case of overcurrent relay co-ordination for Dy transformers the distribution of currents in these transformers should be checked for three-phase, phase-to-phase, and single-phase faults on the secondary winding, shown in Figure 5.18.

To simplify the operations, it can be assumed that the voltages between the phases of the transformer are the same, for both the primary and the secondary windings.

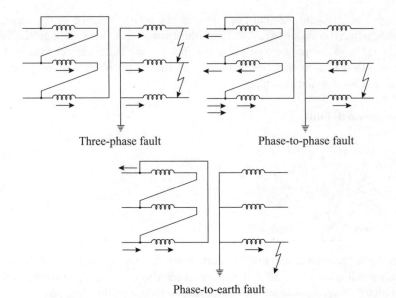

Three-phase fault   Phase-to-phase fault

Phase-to-earth fault

Figure 5.18   Distribution of current for a fault on a Dy transformer

Thus, the number of turns on the primary is equal to $\sqrt{3}$ times the number of turns of the secondary, i.e. $N_1 = \sqrt{3} N_2$.

Three-phase fault

$$I_f = \frac{E_{\phi-n}}{X} = I \tag{5.8}$$

$$I_{\text{delta}} = I \frac{N_2}{N_1} = \frac{I}{\sqrt{3}} \tag{5.9}$$

$$I_{\text{primary}} = \sqrt{3} I_{\text{delta}} = I \tag{5.10}$$

From the above it can be seen that the currents that flow through the relays associated with the secondary winding are equal to the currents that flow through those relays associated with the primary winding, as expected, since the primary and secondary voltages are equal and the fault involves all three phases.

Phase-to-phase fault

$$I_f = \frac{E_{\phi-\phi}}{2X} = \frac{\sqrt{3} \times E_{\phi-n}}{2X} = \frac{\sqrt{3}}{2} I \tag{5.11}$$

$$I_{\text{delta}} = \frac{\sqrt{3}}{2} \times I \times \frac{N_2}{N_1} = \frac{I}{2} \tag{5.12}$$

$$I_{\text{primary}} = 2 I_{\text{delta}} = I \tag{5.13}$$

For this case, the current that goes through the relays installed in the secondary winding circuit is equal to $\sqrt{3}/2$ times the current that flows through the relays associated with the primary on the phase that has the largest current value. From Figure 5.18 it is clear that, for this fault, the current distribution at the primary is 1-1-2, and 0-1-1 at the secondary.

Phase-to-earth fault

$$I_f = \frac{E_{\phi-n}}{X} = I \tag{5.14}$$

$$I_{delta} = I \times \frac{N_2}{N_1} = \frac{I}{\sqrt{3}} \tag{5.15}$$

$$I_{primary} = \frac{I}{\sqrt{3}} \tag{5.16}$$

Thus, for a phase-to-earth fault, the current through the relays installed in the secondary winding circuit on the faulted phase is equal to $\sqrt{3}$ times the current that flows through the relays associated with the primary winding on the same phase.

The results of the three cases are summarised in Table 5.4. Analysing the results, it can be seen that the critical case for the co-ordination of overcurrent relays is the phase-to-phase fault. In this case the relays installed in the secondary carry a current less than the equivalent current flowing through the primary relays, which could lead to a situation where the selectivity between the two relays is at risk. For this reason, the discrimination margin between the relays is based on the operating time of the secondary relays at a current equal to $\sqrt{3}I_f/2$, and on the operating time for the primary relays for the full fault current value I_f, as shown in Figure 5.19.

Example 5.3

For the system shown in Figure 5.20, calculate the following:

1. The three-phase short-circuit levels on busbars 1 and 2.
2. The transformation ratios of the CTs associated with breakers 1 to 8, given that the primary turns are multiples of 100 except for the CT for breaker number 9, which has a ratio of 250/5. Assume that the total burden connected to each CT is 1 Ω and that C100 cores are used.

Table 5.4   Summary of fault conditions

Fault	$I_{primary}$	$I_{secondary}$
Three-phase	I	I
Phase-to-phase	I	$\sqrt{3}I/2$
Phase-to-earth	I	$\sqrt{3}I$

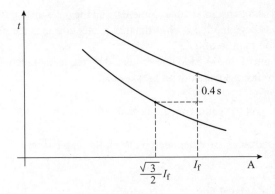

Figure 5.19   Co-ordination of overcurrent relays for a Dy transformer

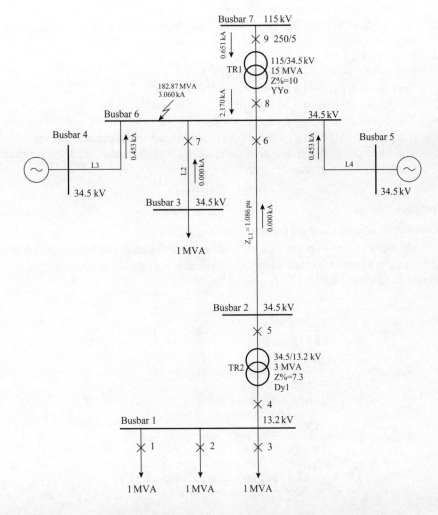

Figure 5.20   Single line diagram for Example 5.3 showing fault currents

3. The settings of the instantaneous elements, and the pick-up and time dial settings of the relays to guarantee a co-ordinated protection arrangement, allowing a discrimination margin of 0.4 s.
4. The percentage of the 34.5 kV line protected by the instantaneous element of the overcurrent relay associated with breaker 6.

Take into account the following additional information:

1. The p.u. impedances are calculated on the following bases:
    $V = 34.5\,\text{kV}$
    $P = 100\,\text{MVA}$
2. The settings of relay 7 are:
    Pick-up $= 4\,\text{A}$
    Time dial $= 0.3$
    Instantaneous $= 1100\,\text{A}$ primary current
3. All the relays have an IEC very inverse time characteristic with the values given in Table 5.1
4. Relay data:
    Pick-up setting $= 1$ to $12\,\text{A}$ in steps of $1\,\text{A}$
    Time dial setting $= 0.05$ to 10 in steps of 0.05
    Instantaneous: 6 to $14\,\text{A}$ in steps of $1\,\text{A}$
5. The setting of the instantaneous elements of the relays associated with the feeders is to be carried out on the basis of ten times the maximum nominal current.
6. The short-circuit MVA and fault currents, for a fault on the 34.5 kV busbar at substation A, are given in Figure 5.20.

Solution

Calculation of equivalent impedance

The short-circuit level on the 34.5 kV busbar at substation A can be obtained from the values in Figure 5.20 (183.11 MVA). Using this, the equivalent impedance of the system behind the busbar is calculated as follows:

$$Z_{base} = \frac{V^2}{P_{sc}} = \frac{(34500)^2}{183.11 \times 10^6} = 6.5\,\Omega,\ \text{referred to } 34.5\,\text{kV}$$

$$Z_{transf1} = 0.1\frac{(34500)^2}{15 \times 10^6} = 7.93\,\Omega,\ \text{referred to } 34.5\,\text{kV}$$

$$= 88.17\,\Omega,\ \text{referred to } 115\,\text{kV}$$

$$Z_{transf\,2} = 0.073\frac{(34500)^2}{3 \times 10^6} = 28.96\,\Omega,\ \text{referred to } 34.5\,\text{kV}$$

$$Z_{line} = 1.086\frac{(34500)^2}{100 \times 10^6} = 12.93\,\Omega,\ \text{referred to } 34.5\,\text{kV}$$

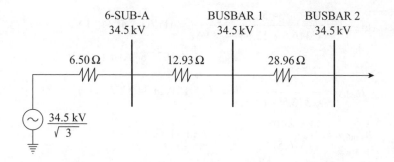

Figure 5.21   Positive sequence network for Example 5.3

The equivalent positive sequence network, referred to 34.5 kV, is shown in Figure 5.21.

Nominal currents

$$I_{\text{nom}1,2,3} = \frac{1 \times 10^6}{\sqrt{3} \times 13.2 \times 10^3} = 43.74\,\text{A at }13.2\,\text{kV}$$

$$I_{\text{nom}4} = \frac{3 \times 10^6}{\sqrt{3} \times 13.2 \times 10^3} = 131.22\,\text{A at }13.2\,\text{kV}$$

$$I_{\text{nom}5} = \frac{3 \times 10^6}{\sqrt{3} \times 34.5 \times 10^3} = 50.20\,\text{A at }34.5\,\text{kV}$$

$$I_{\text{nom}6} = 50.20\,\text{A at }34.5\,\text{kV}$$

$$I_{\text{nom}7} = \frac{1 \times 10^6}{\sqrt{3} \times 34.5 \times 10^3} = 16.73\,\text{A}$$

$$I_{\text{nom}8} = 251.02\,\text{A at }34.5\,\text{kV}$$

$$I_{\text{nom}9} = \frac{15 \times 10^6}{\sqrt{3} \times 115 \times 10^3} = 75.31\,\text{A at }115\,\text{kV}$$

Short-circuit levels

The short-circuit MVA (P_{sc}) of the transformer at 34.5 kV $= \sqrt{3} \times 2170.34 \times 34.5 \times 10^3 = 129.69\,\text{MVA}$.

$$Z_{\text{transf1}} + Z_{\text{base}} = \frac{(34.5)^2}{129.69} = 9.18\,\Omega,\text{ referred to }34.5\,\text{kV}$$

$$= 101.97\,\Omega,\text{ referred to }115\,\text{kV}$$

$$Z_{\text{system}} = 101.97\,\Omega - 88.17\,\Omega = 13.80\,\Omega,\text{ referred to }115\,\text{kV}$$

$$I_{\text{fault}1,2,3,4} = \frac{34.5 \times 10^3}{\sqrt{3}\,(6.5 + 12.93 + 28.96)} = 411.63 \text{ A at } 34.5 \text{ kV},$$

$$= 1075.84 \text{ A at } 13.2 \text{ kV}$$

$$I_{\text{fault}5} = \frac{34.5 \times 10^3}{\sqrt{3}\,(6.5 + 12.93)} = 1025.15 \text{ A at } 34.5 \text{ kV}$$

$$I_{\text{fault}6,7} = \frac{34.5 \times 10^3}{\sqrt{3}\,(6.5)} = 3064.40 \text{ A at } 34.5 \text{ kV}$$

$$I_{\text{fault}8} = \frac{129.69 \times 10^6}{\sqrt{3} \times 34.5 \times 10^3} = 2170.34 \text{ A at } 34.5 \text{ kV}$$

$$I_{\text{fault}9} = \frac{115 \times 10^3}{\sqrt{3} \times 13.80} = 4811.25 \text{ A at } 115 \text{ kV}$$

Selection of current transformers

Table 5.5 gives the main values for determining the transformation ratio of the CTs, which is taken as the larger of the two following values:

- nominal current;
- maximum short-circuit current for which no saturation is present.

Therefore, $(I_{\text{sc}} \times 5/X) \leq 100 \Rightarrow X \geq (I_{\text{sc}} \times 5/100)$

Determining the pick-up (PU) values

$$I_{\text{load}1,2,3} = 43.74 \text{ A}; \ PU_{1,2,3} = (1.5)(43.74)(5/100) = 3.28 \text{ A}; \ PU_{1,2,3} = 4 \text{ A}$$

$$I_{\text{load}4} = 131.22 \text{ A}; \ PU_4 = (1.5)(131.22)(5/200) = 4.92 \text{ A}; \ PU_4 = 5 \text{ A}$$

$$I_{\text{load}5} = 50.20 \text{ A}; \ PU_5 = (1.5)(50.20)(5/100) = 3.76 \text{ A}; \ PU_5 = 4 \text{ A}$$

Table 5.5   Determination of CT ratios for Example 5.3

Breaker number	P_{nom} (MVA)	I_{nom} (A)	I_{sc} (A)	$(5/100)I_{\text{sc}}$ (A)	CT ratio
9	15	75.31	4797.35	239.87	250/5
8	15	251.02	2170.40	108.51	300/5
7	1	16.73	3060.34	153.01	200/5
6	3	50.20	3060.34	153.01	200/5
5	3	50.20	1025.67	51.28	100/5
4	3	131.22	1076.06	53.80	200/5
1, 2, 3	1	43.74	1076.06	53.80	100/5

$I_{load6} = 50.20$ A; $PU_6 = (1.5)(50.20)(5/200) = 1.88$ A; $PU_6 = 2$ A

$PU_7 = 4$ A (as given in the example data)

$I_{load8} = 251.02$ A; $PU_8 = (1.5)(251.02)(5/300) = 6.28$ A; $PU_8 = 7$ A

$I_{load9} = 75.31$ A; $PU_9 = (1.5)(75.31)(5/250) = 2.26$ A; $PU_9 = 3$ A

Determining the instantaneous and time dial settings

Relays 1, 2 and 3

When calculating the settings for the relays situated at the end of the circuit, the minimum time dial setting of 0.05 is selected. From the given information, the setting of the instantaneous element is based on ten times the maximum load current seen by the relay. Thus, $I_{inst.\ trip} = 10 \times I_{nom} \times (1/CTR) = 10 \times 43.74 \times (5/100) = 21.87$ A $\Rightarrow$ set at 22 A. $I_{prim.\ trip} = 22(100/5) = 440$ A.

Given the constants for the IEC very inverse overcurrent relay are $\alpha = 1.0$, $\beta = 13.5$, and $L = 0$ then, from eqn. 5.7, the relay operating time t is [(time dial setting) $\times 13.5]/(PSM - 1)$, where PSM is the ratio of the fault current in secondary amps to the relay pick-up current. $PSM = 22/4 = 5.5$ times and, with the time dial setting of 0.05, the relay operating time is $(0.05 \times 13.5)/(5.5 - 1) = 0.15$ s.

Relay 4

To discriminate with relay 3 at 440 A requires operation in $t_{4a} - 0.15 + 0.4 = 0.55$ s.

$PSM_{4a} = (440 \times 5/200)(1/5) = 2.2$ times. At 2.2 times, and $t_{4a} = 0.55$ s, the time dial setting $= 0.55 \times (2.2 - 1)/13.5 = 0.049 \Rightarrow 0.05$.

This relay has no setting for the instantaneous element, as referred to in Section 5.3.1. The operating time for a line-to-line fault is determined by taking 86 per cent of the three-phase fault current. $PSM_{4b} = (0.86)(1075.84 \times 5/200)(1/5) = 4.63$ times. By similar calculations to those for relays 1, 2 and 3, $t_{4b} = 0.19$ s.

Relay 5

The back-up to relay 4 is obtained by considering the operating time for a line-to-line fault of $t_{5a} = 0.19 + 0.4 = 0.59$ s.

$PSM_{5a} = 1075.84 \times (13.2/34.5) \times (5/100) \times (1/4) = 5.15$ times. At 5.15 times, and $t_{5a} = 0.59$ s, this gives a required time dial setting of 0.20.

The setting of the instantaneous element is $1.25(1075.84) \times (13.2/34.5) \times (5/100) = 25.73$ A $\Rightarrow 26$ A, so that $I_{prim.\ trip} = 26(100/5) = 520$ A.

The operating time of the time delay element is calculated from $PSM_{5b} = (1/4) \times 26 = 6.5$ times. At 6.5 times, and with a time dial setting of 0.20, from the relay characteristics and eqn. 5.7, $t_{5b} = 0.5$ s.

Relay 6

At 520 A, this relay has to operate in $t_{6a} = 0.5 + 0.4 = 0.9$ s.

$PSM_{6a} = 520 \times (5/200) \times (1/2) = 6.5$ times. At 6.5 times and $t_{6a} = 0.9$ s, the time dial setting $= 0.37 \Rightarrow 0.40$.

Instantaneous setting $= 1.25(1025.15)(5/200) = 32.04 \Rightarrow 32$ A. $I_{prim.\ trip} = 32$ A $\times 200/5 = 1280$ A.

Relay 7: PSM and DIAL as given in the example data.

Relay 8

Relay backs up relays 6 and 7 and should be co-ordinated with the slower of these two relays. Relay 7 has an instantaneous primary current setting of 1100 A, equivalent to 27.5 A secondary current which is less than the setting of relay 6, and so the operating time of both relays is determined by this value.

For relay 7 $PSM = 1100 \times 5/200 \times 1/4 = 6.87$ times. At 6.87 times and with a time dial setting of 0.3, then $t_{op} = 0.69$ s.

For relay 6 $PSM = 1100 \times 5/200 \times 1/2 = 13.75$ times. At 13.75 times and with a time dial setting of 0.4, $t_{op} = 0.42$ s.

Therefore the operating time to give correct discrimination with relay 7 is $t_{8a} = 0.69 + 0.4 = 1.09$ s.

For back-up relay 8, the contributions to relay 6 from substations G and M are not considered. Only the infeed from the transformer has to be taken into account, so that $PSM_{8a} = 1100 \times (2170.34/3060.40) \times (5/300) \times (1/7) = 1.86$ times. At 1.86 times and $t_{8a} = 1.09$ s, the time dial setting $= 0.07 \Rightarrow 0.1$.

Here also, no instantaneous setting is applied to relay 8 for the reasons given in Section 5.2.1. The maximum short-circuit current to be used for this relay is that which flows from the 115 kV busbar to the 34.5 kV busbar for a fault on the latter, and $PSM_{8b} = 2170.34 \times (5/300) \times (1/7) = 5.17$ times. At 5.17 times and with a time dial setting of 0.1, $t_{8b} = 0.32$ s.

Relay 9

This relay backs up relay 8 in a time of $t_{9a} = 0.4 + 0.32 = 0.72$ s.

$PSM_{9a} = 2170.34 \times (34.5/115) \times (5/250) \times (1/3) = 4.34$ times. At 4.34 times and $t_{9a} = 0.72$ s, the time dial setting $= 0.18 \Rightarrow 0.20$.

The instantaneous setting $= 1.25 \times 2170.39 \times (34.5/115) \times (5/250) = 16.28$ A $\Rightarrow$ 17 A. $I_{prim.\ trip} = 17 \times (250/5) = 850$ A referred to 115 kV.

The co-ordination curves of the relays associated with this system are shown in Figure 5.22, and summarised in Table 5.6.

Percentage of 34.5 kV line protected by the instantaneous element of the overcurrent relay associated with breaker 6

Given

$$\% = \frac{K_s(1 - K_i) - 1}{K_i}$$

where

$$K_i = \frac{I_{sc.pickup}}{I_{sc.end}} = \frac{1280}{1025.15} = 1.25$$

and

$$K_s = \frac{Z_{source}}{Z_{element}}$$

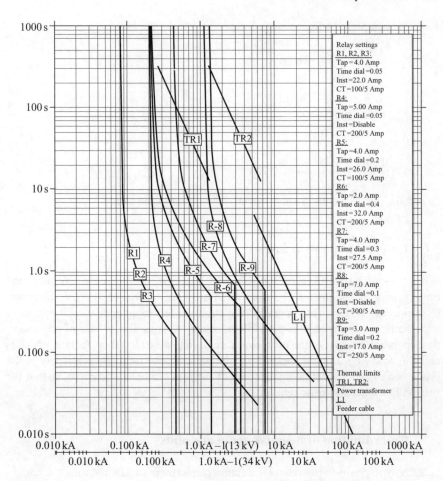

Figure 5.22   Relay co-ordination curves for Example 5.3

Table 5.6   Summary of settings for Example 5.3

Relay number	CT ratio	Pick-up (A)	Time dial	Instantaneous I_{sec}(A)
1, 2, 3	100/5	4	1/2	20.0
4	200/5	5	1/2	–
5	100/5	4	3	26.0
6	200/5	2	6	32.0
7	200/5	4	5	27.5
8	300/5	7	1	–
9	250/5	3	2	17.0

From the computer listing

$$Z_f = \frac{V^2}{P} = \frac{34.5^2}{183.11} = 6.50$$

and

$$K_s = 6.50/12.93 = 0.50$$

Therefore

$$\% = \frac{0.50(1 - 1.25) + 1}{1.25} = 0.70$$

so that the instantaneous element covers 70 per cent of the line.

5.6   Co-ordination with fuses

When co-ordinating overcurrent relays it may be necessary to consider the time/current characteristics of fuses which are used to protect MV/LV substation transformers. When a fuse operates, the circuit is left in an open-circuit condition until the fuse is replaced. It is therefore necessary to consider the case of preventing fuse operation because of the problems of replacing them after they operate, which is called fuse saving. In these cases it may be preferable to forgo the selectivity of the protection system by not taking account of the fuse characteristic curve, so that the fuse will then act as a back-up.

5.7   Co-ordination of negative-sequence units

Sensitivity to phase-to-phase fault detection can be increased by the use of negative-sequence relays (type 50/51 Q) because a balanced load has no negative-sequence (I_2) current component. This is also the situation for phase-to-earth faults if type 50/51 relays are used since a balanced load has no zero-sequence (I_0) component.

Instantaneous overcurrent and time-delay overcurrent negative-sequence units are common features of the new multifunction relays. It is important to ensure that the settings of these units are checked for co-ordination with phase-only sensing devices such as downstream fuses and reclosers, and/or earth-fault relays.

To determine the pick-up setting of the negative-sequence element, it is necessary to keep in mind that the magnitude of the current for a phase-to-phase fault is $\sqrt{3}/2$ (87 per cent) of the current for a three-phase fault at the same location, as indicated in eqn. 5.11. On the other hand, the magnitude of the negative-sequence component for a phase-to-phase fault can be obtained from the following expression taken from Section 2.2: $I_{a2} = 1/3(I_a + a^2 I_b + a I_c)$. For a phase-to-phase fault, $I_a = 0$ and $I_b = -I_c$. Therefore the magnitude of the negative-sequence component is $1/\sqrt{3}$ (58 per cent) of the magnitude of the phase-fault current. When the two factors ($\sqrt{3}/2$ and $1/\sqrt{3}$) are combined, the $\sqrt{3}$ factors cancel, leaving a 1/2 factor.

From the above it is recommended that negative-sequence elements be set by taking one half of the phase pick-up setting in order to achieve equal sensitivity to phase-to-phase faults as to three-phase faults.

To plot the negative-sequence time/current characteristics on the same diagram as the phase- and earth-fault devices it is necessary to adjust the negative-sequence element pick-up value by a multiplier that is the ratio of the fault current to the negative-sequence current. For a phase-to-phase fault this is 1.732. The negative-sequence pick-up value should be multiplied by a value greater than 1.732 for a phase-to-phase-to-earth fault, and by a factor of 3 for a phase-to-earth fault. Since no negative-sequence current flows for a three-phase fault, negative-sequence relay operation does not take place, and no multiplying factor is involved.

Consider a downstream time-delay phase overcurrent element with a pick-up value of 100 A, and an upstream negative-sequence time-delay relay with a pick-up value of 150 A. In order to check the co-ordination between these two units for a phase-to-phase fault, the phase overcurrent element plot should be shifted to the right by a factor of 1.732, with a pick-up value of $1.732 \times 150 = 259.8$ A. Generally, for co-ordination with downstream phase overcurrent devices, phase-to-phase faults are the most critical with all other fault types resulting in an equal or greater shift of the time/current characteristic curve to the right on the co-ordination graph.

5.8   Overcurrent relays with voltage control

Faults close to generator terminals may result in voltage drop and fault current reduction, especially if the generators are isolated and the faults are severe. Therefore, in generation protection it is important to have voltage control on the overcurrent time-delay units to ensure proper operation and co-ordination. These devices are used to improve the reliability of the relay by ensuring that it operates before the generator current becomes too low. There are two types of overcurrent relays with this feature – voltage-controlled and voltage-restrained, which are generally referred to as type 51 V relays.

The voltage-controlled (51/27C) feature allows the relays to be set below rated current, and operation is blocked until the voltage falls well below normal voltage. The voltage-controlled approach typically inhibits operation until the voltage drops below a pre-set value. It should be set to function below about 80 per cent of rated voltage with a current pick-up of about 50 per cent of generator rated current.

The voltage-restrained (51/27R) feature causes the pick-up to decrease with reducing voltage, as shown in Figure 5.23. For example, the relay can be set for 175 per cent of generator rated current with rated voltage applied. At 25 per cent voltage the relay picks up at 25 per cent of the relay setting ($1.75 \times 0.25 = 0.44$ times rated). The varying pick-up level makes it more difficult to co-ordinate the relay with other fixed pick-up overcurrent relays.

Since the voltage-controlled type has a fixed pick-up, it can be more readily co-ordinated with external relays than the voltage-restrained type. On the other hand, compared to the voltage-controlled relay, the voltage-restrained type will be less

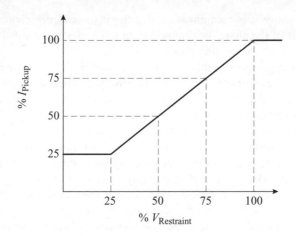

Figure 5.23   Pick-up setting of 51/27R relay

susceptible to operation on swings or motor-starting conditions that depress the voltage below the voltage-controlled undervoltage unit drop-out point.

5.9   Setting overcurrent relays using software techniques

The procedure for determining the settings of overcurrent relays, as illustrated in the past sections, is relatively simple for radial or medium-sized interconnected systems. However, for large systems, the procedure becomes cumbersome if performed manually and therefore software techniques are required, especially if different topologies have to be analysed. This section introduces a very simple procedure to set overcurrent relays using different algorithms. The entry data required are the short-circuit currents for faults at all busbars, the margins and constraints of the system and the available settings of the relays being co-ordinated. In addition, the settings of those relays closest to the loads and at the boundaries of other networks have to be considered.

The overall process consists basically of three steps:

1. Locate the fault and obtain the current for setting the relays.
2. Identify the pairs of relays to be set, determining first which one is furthest away from the source, and which is acting as the back-up. The program should define the settings in accordance with the criteria given in Section 5.3.
3. Verify that the requirements given in Section 5.4 are fulfilled; otherwise the process should be repeated with lower discrimination margins, or new relays should be tried.

The single line diagram given in Figure 5.20 can be used to illustrate simply how a typical computer program can tackle a co-ordination problem.

The algorithm files the discrimination margins required, the setting of the relays closest to the loads, (1, 2 and 3 in this case), and those for relay 9 which corresponds to the only boundary utility. The algorithm then establishes pairs of relays and identifies which relay acts as a back-up within each pair. For the system shown in Figure 5.20, the algorithm will determine which is the slowest relay of 1, 2 and 3, and co-ordinate this with relay 4 located upstream. Relay 4 will, in turn, be co-ordinated with relay 5, and relay 5 with relay 6, by ensuring that the required time discrimination margin is maintained in all cases. Similar procedures are carried out for those relays associated with the rest of the lines connected to busbar 6. After this, the algorithm determines the slowest of these relays which then has to be co-ordinated with relay 8, and finally relay 8 is co-ordinated with relay 9. When the process is finished, the algorithm carries out all the necessary checks in accordance with the restraints given in the data entry. If any requirement is not fulfilled, the process is started again with a lower discrimination margin or using relays with different characteristics until adequate co-ordination is achieved.

During the execution of the program the critical route, which corresponds to the one with the highest number of relay pairs, should be identified. The more inter-connected the system, the larger and more complicated will be the critical route and computer programs are being used more and more for large systems. However, for small systems and fault-case analysis, manual methods are still used with the help of software editors containing libraries with relay curves from many manufacturers. This reduces the curve drawing process, which is very time consuming.

5.10   Use of digital logic in numerical relaying

5.10.1   General

When using numerical relays it is necessary to provide a suitable method for handling the relay logic capabilities, which should include blocks with control inputs, virtual outputs, and hardware outputs. A group of logic equations defining the function of the multifunction relay is called a logic scheme.

Numerical relays can be configured to suit a particular specification by defining the operating settings (pick-up thresholds and time delays) of the individual protection and control functions. Operating settings and logic settings are interdependent, but separately programmed functions. Changing logic settings is similar to rewiring a panel, and is used for managing the input, output, protection, control, monitoring and reporting capabilities of multifunction protection relay systems. Each relay system has multiple, normally-contained function blocks that have all of the inputs and outputs of its discrete component counterpart. Each independent function block interacts with control inputs, virtual outputs, and hardware outputs based on logic variables defined in equation form with relay logic. Relay logic equations entered and saved in the relay system's nonvolatile memory integrate the selected or enabled protection in order to provide the operating settings that control the relay pick-up threshold and time delay values.

5.10.2   Principles of digital logic

Digital systems are constructed by using three basic logic gates. These gates are designated AND, OR and NOT. There also exist other logical gates, such as the

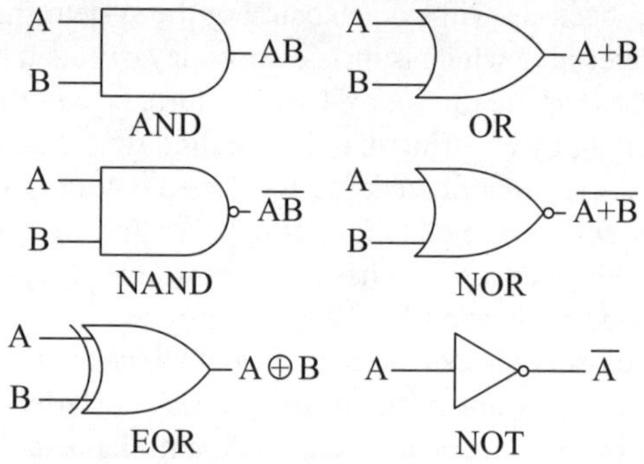

Figure 5.24   Logic gate symbols

NAND and EOR gates. The basic operations and logic gate symbols are summarised in Figure 5.24.

The AND gate is an electronic circuit that gives a high output only if all its inputs are high. A dot (.) is used to show the AND operation but this dot is usually omitted, as in Figure 5.24. The OR gate gives a high output if one or more of its inputs are high. A plus sign (+) is used to show the OR operation. The NOT gate produces an inverted version of the input at its output. It is also known as an inverter. If the variable is A, the inverted output is known as NOT A and is shown as $\overline{A}$. The NAND is a NOT-AND circuit that is equal to an AND circuit followed by a NOT circuit. The outputs of all NAND circuits are high if any of the inputs are low. The NOR gate is a NOT-OR circuit that is equal to an OR circuit followed by a NOT circuit. The outputs of all NOR gates are low if any of the inputs are high. The EOR – the Exclusive OR gate – is a circuit that will give a high output if either, but not both, of its two inputs are high. An encircled plus sign, $\oplus$, is used to show the EOR operation.

Table 5.7 shows the input/output combinations for the gate functions mentioned above. Note that a truth table with n inputs has $2n$ rows.

5.10.3   Logic schemes

Normally numerical relays have several pre-programmed logic schemes which are stored in the relay memory. Each scheme is configured for a typical protection application and virtually eliminates the need for start-from-scratch programming. Protection scheme designers may select from a number of pre-programmed logic schemes that perform the most common protection and control requirements. Alternatively customised schemes can be created using the relay logic capabilities.

Figure 5.25 shows a typical pre-programmed logic scheme provided with a numerical relay, where the features of each logic scheme are broken down into functional groups. This logic scheme provides basic time and instantaneous overcurrent protection. The protective elements include phase, neutral, and negative-sequence overcurrent protection. Functions such as breaker failure, virtual breaker control,

*Table 5.7a   Input/output combinations for
various gate functions*

Inputs		Outputs				
A	B	AND	OR	NAND	NOR	EOR
0	0	0	0	1	1	0
0	1	0	1	1	0	1
1	0	0	1	1	0	1
1	1	1	1	0	0	0

*Table 5.7b   Input/output
combinations for
the NOT gate*

Input	Output
0	1
1	0

automatic reclosing and protective voltage features are not enabled in this scheme. However, these features may be achieved by appropriate design of the relay logic.

This numerical relay has 4 programmable inputs IN1 to IN4; five programmable outputs OUT1 to OUT5; one programmable alarm output OUTA; ten virtual outputs VO6 to VO15; four virtual selection switches 43, 143, 243, 343 and four protection setting groups with external or automatic selection modes.

The phase, neutral and negative-sequence elements are activated to provide timed (51) and instantaneous (50) overcurrent protection in this scheme. A function block is disabled by setting the pick-up set-point at zero in each of the four setting groups. Virtual output VO11 is assigned for all protective trips. When VO11 becomes TRUE, OUT1 will operate and trip the breaker. Contact outputs OUT2, OUT3, OUT4, and OUT5 are designated to specific function blocks. OUT2 operates for instantaneous phase overcurrent conditions, OUT3 trips for timed phase overcurrent situations, OUT4 operates for instantaneous neutral and negative-sequence overcurrent conditions, and OUT5 operates for timed neutral and negative-sequence overcurrent conditions. Input 1 IN1 is typically assigned to monitor breaker status (52b). A setting group can be selected automatically or by using the communication ports or the front panel HMI. Automatic setting group changes are normally based on current level and duration. Setting group changes initiated by contact sensing inputs are not accommodated in this scheme, but the logic inputs can be programmed to provide this function.

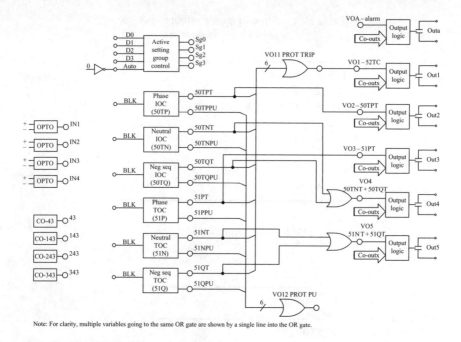

Figure 5.25   Typical pre-programmed logic scheme of a numerical relay (reproduced by permission of Basler Electric)

5.11   Adaptive protection with group settings change

This section presents several examples of logic scheme customisation to provide functions that normally are not incorporated in numerical relays as part of a manufacturer's package. Topology changes, for example, affect the short-circuit levels and therefore an incorrect co-ordination might arise if the relays are not reset for the prevailing power system conditions. To overcome this, adaptive protection, which can be implemented by using the multiple setting groups feature included in most numerical relays, is essential.

Figure 5.26 shows a portion of a power system that might have four scenarios as follows:

- system normal;
- one of the transformers out of service for maintenance;
- the grid infeed not available;
- the in-house generator out of service.

For a fault on one of the feeders, resulting in a fault current of I_f, from the co-ordination curve R_1 in Figure 5.27 the feeder relay would operate in a time t_1. With both transformers in service the fault current passing through each transformer would

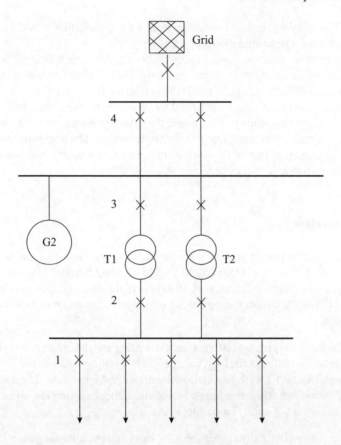

Figure 5.26   Electrical system to illustrate setting group change

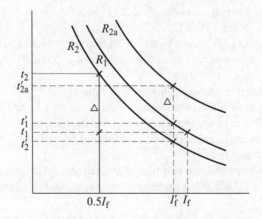

Figure 5.27   Overcurrent co-ordination curves considering adaptive relaying

be $0.5I_f$. The relay R_2 on the low voltage side of the transformer would then operate in time t_2, giving a discrimination margin of $(t_2 - t_1)$.

However, when one transformer is out of service, the current through the remaining transformer increases to I_f'. From curve R_2, the transformer relay tripping time would then reduce from t_2 to t_2', which is faster than the feeder relay operating time t_1', leading to incorrect relay operation. It is thus necessary to move the transformer low voltage relay curve upwards to R_{2a}, where the relay operating time at I_f' will be t_{2a}', in order to maintain discrimination with the feeder relays. The transformer relays can be pre-programmed to ensure this shift of the relay curve takes place automatically when one transformer is out of service.

5.12   Exercises

5.1   Consider a power system with the same single line diagram as used for Example 5.2, but with a 115/34.5 kV 58.5 MVA transformer. The feeders from substation C each have a capacity of 10 MVA. If the short-circuit level at substation A is 1400 MVA, determine the setting of relays 1, 2, 3 and 4 if the same type of relay is used.

5.2   Calculate the pick-up setting, time dial setting and the instantaneous setting of the phase relays installed in the high voltage and low voltage sides of the 115/13.2 kV transformers T1 and T3, in the substation illustrated in Figure 5.28. The short-circuit levels, CT ratios and other data are shown in the same diagram. The considerations used for Examples 5.2 and 5.3 also apply here.

5.3   For the system shown in Figure 5.29, carry out the following calculations:

1. The maximum values of short-circuit current for three-phase faults at busbars A, B and C, taking into account that busbar D has a fault level of 12906.89 A r.m.s. symmetrical (2570.87 MVA).
2. a) The maximum peak values to which breakers 1, 5 and 8 can be subjected.
    b) The r.m.s. asymmetrical values that breakers 1, 5 and 8 can withstand for 5 cycles for guarantee purposes.
    For these calculations assume that the L/R ratio is 0.2.
3. The turns ratios of the CTs associated with breakers 1 to 8. The CT in breaker 6 is 100/5. Take into account that the secondaries are rated at 5 A and that the ratios available in the primaries are multiples of 50 up to 400, and from then on are in multiples of 100.
4. The instantaneous, pick-up and time dial settings for the phase relays in order to guarantee a co-ordinated protection system, allowing a time discrimination margin of 0.4 s.
5. The percentage of the 34.5 kV line that is protected by the instantaneous element of the overcurrent relay associated with breaker 5.

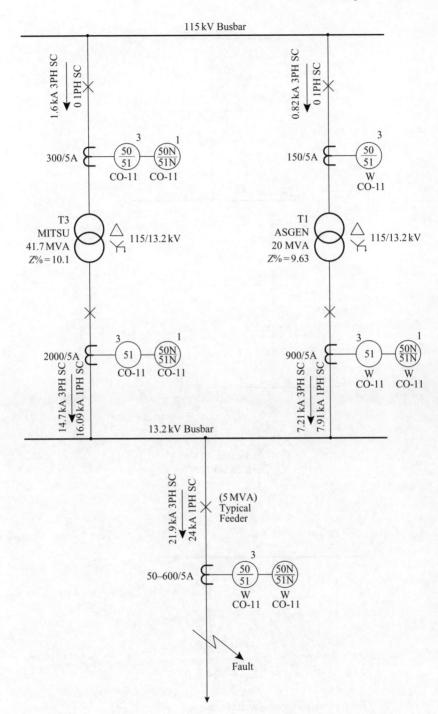

Figure 5.28   Single line diagram for Exercise 5.2

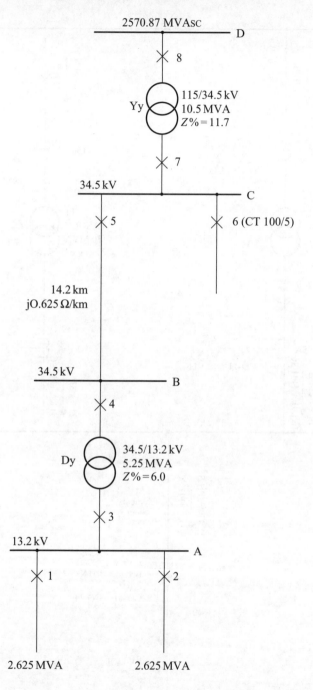

Figure 5.29   Single line diagram for Exercise 5.3

Bear in mind the following additional information:

- The settings of relay 6 are as follows: pick-up 7 A, time dial 5, instantaneous setting 1000 A primary current.
- All the relays are inverse time type, with the following characteristics:
  Pick-up: 1 to 12 in steps of 2 A
  Time dial: as in Figure 5.15
  Instantaneous element: 6 to 144 in steps of 1 A

Calculate the setting of the instantaneous elements of the relays associated with the feeders assuming 0.5 I_{sc} on busbar A.

Chapter 6

Fuses, reclosers and sectionalisers

A wide variety of equipment is used to protect distribution networks. The particular type of protection used depends on the system element being protected and the system voltage level, and, even though there are no specific standards for the overall protection of distribution networks, some general indication of how these systems work can be made.

6.1 Equipment

The devices most used for distribution system protection are:

- overcurrent relays;
- reclosers;
- sectionalisers;
- fuses.

The co-ordination of overcurrent relays was dealt with in detail in the previous chapter, and this chapter will cover the other three devices referred to above.

6.1.1 Reclosers

A recloser is a device with the ability to detect phase and phase-to-earth overcurrent conditions, to interrupt the circuit if the overcurrent persists after a predetermined time, and then to automatically reclose to re-energise the line. If the fault that originated the operation still exists, then the recloser will stay open after a preset number of operations, thus isolating the faulted section from the rest of the system. In an overhead distribution system between 80 to 95 per cent of the faults are of a temporary nature and last, at the most, for a few cycles or seconds. Thus, the recloser, with its opening/closing characteristic, prevents a distribution circuit being left out of service for temporary faults. Typically, reclosers are designed to have up to three open-close operations and, after these, a final open operation to lock out the sequence.

One further closing operation by manual means is usually allowed. The counting mechanisms register operations of the phase or earth-fault units which can also be initiated by externally controlled devices when appropriate communication means are available.

The operating time/current characteristic curves of reclosers normally incorporate three curves, one fast and two delayed, designated as A, B and C, respectively. Figure 6.1 shows a typical set of time/current curves for reclosers. However, new reclosers with microprocessor-based controls may have keyboard-selectable

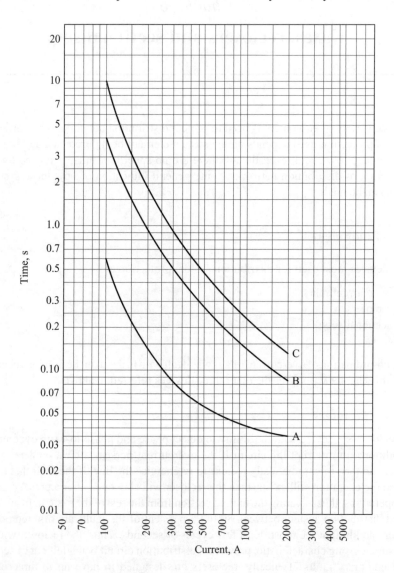

Figure 6.1   Time/current curves for reclosers

time/current curves which enable an engineer to produce any curve to suit the co-ordination requirements for both phase and earth-faults. This allows reprogramming of the characteristics to tailor an arrangement to a customer's specific needs without the need to change components.

Co-ordination with other protection devices is important in order to ensure that, when a fault occurs, the smallest section of the circuit is disconnected to minimise disruption of supplies to customers. Generally, the time characteristic and the sequence of operation of the recloser are selected to co-ordinate with mechanisms upstream towards the source. After selecting the size and sequence of operation of the recloser, the devices downstream are adjusted in order to achieve correct co-ordination. A typical sequence of a recloser operation for a permanent fault is shown in Figure 6.2. The first shot is carried out in instantaneous mode to clear temporary faults before they cause damage to the lines. The three later ones operate in a timed manner with predetermined time settings. If the fault is permanent, the time-delay operation allows other protection devices nearer to the fault to open, limiting the amount of the network being disconnected.

Earth faults are less severe than phase faults and, therefore, it is important that the recloser has an appropriate sensitivity to detect them. One method is to use CTs connected residually so that the resultant residual current under normal conditions is approximately zero. The recloser should operate when the residual current exceeds the setting value, as would occur during earth faults.

Reclosers can be classified as follows:

- single-phase and three-phase;
- mechanisms with hydraulic or electronic operation;
- oil, vacuum or SF_6.

Single-phase reclosers are used when the load is predominantly single-phase. In such a case, when a single-phase fault occurs the recloser should permanently disconnect the faulted phase so that supplies are maintained on the other phases. Three-phase reclosers are used when it is necessary to disconnect all three phases in order to prevent unbalanced loading on the system.

Reclosers with hydraulic operating mechanisms have a disconnecting coil in series with the line. When the current exceeds the setting value, the coil attracts a piston that opens the recloser main contacts and interrupts the circuit. The time characteristic and operating sequence of the recloser are dependent on the flow of oil in different chambers. The electronic type of control mechanism is normally located outside the recloser and receives current signals from a CT-type bushing. When the current exceeds the predetermined setting, a delayed shot is initiated that finally results in a tripping signal being transmitted to the recloser control mechanism. The control circuit determines the subsequent opening and closing of the mechanism, depending on its setting. Reclosers with electronic operating mechanisms use a coil or motor mechanism to close the contacts. Oil reclosers use the oil to extinguish the arc and also to act as the basic insulation. The same oil can be used in the control mechanism. Vacuum and SF_6 reclosers have the advantage of requiring less maintenance.

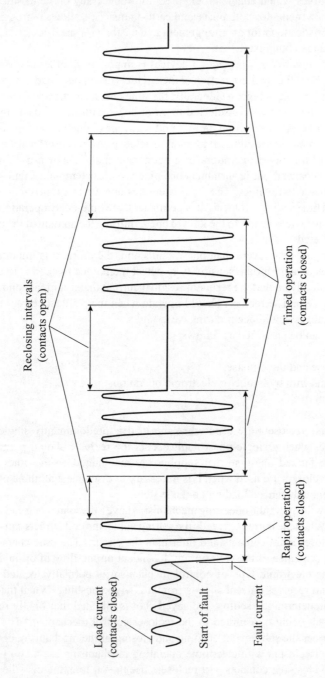

Figure 6.2   Typical sequence for recloser operation

Reclosers are used at the following points on a distribution network:

- in substations, to provide primary protection for a circuit;
- in main feeder circuits, in order to permit the sectioning of long lines and thus prevent the loss of a complete circuit due to a fault towards the end of the circuit;
- in branches or spurs, to prevent the tripping of the main circuit due to faults on the spurs.

When installing reclosers it is necessary to take into account the following factors:

1. System voltage.
2. Short-circuit level.
3. Maximum load current.
4. Minimum short-circuit current within the zone to be protected by the recloser.
5. Co-ordination with other mechanisms located upstream towards the source, and downstream towards the load.
6. Sensitivity of operation for earth faults.

The voltage rating and the short-circuit capacity of the recloser should be equal to, or greater than, the values that exist at the point of installation. The same criteria should be applied to the current capability of the recloser in respect of the maximum load current to be carried by the circuit. It is also necessary to ensure that the fault current at the end of the line being protected is high enough to cause operation of the recloser.

6.1.2 Sectionalisers

A sectionaliser is a device that automatically isolates faulted sections of a distribution circuit once an upstream breaker or recloser has interrupted the fault current and is usually installed downstream of a recloser. Since sectionalisers have no capacity to break fault current, they must be used with a back-up device that has fault current breaking capacity. Sectionalisers count the number of operations of the recloser during fault conditions. After a preselected number of recloser openings, and while the recloser is open, the sectionaliser opens and isolates the faulty section of line. This permits the recloser to close and re-establish supplies to those areas free of faults. If the fault is temporary, the operating mechanism of the sectionaliser is reset.

Sectionalisers are constructed in single- or three-phase arrangements with hydraulic or electronic operating mechanisms. A sectionaliser does not have a current/time operating characteristic, and can be used between two protective devices whose operating curves are very close and where an additional step in co-ordination is not practicable.

Sectionalisers with hydraulic operating mechanisms have an operating coil in series with the line. Each time an overcurrent occurs the coil drives a piston that activates a counting mechanism when the circuit is opened and the current is zero by the displacement of oil across the chambers of the sectionaliser. After a pre-arranged number of circuit openings, the sectionaliser contacts are opened by means of pretensioned springs. This type of sectionaliser can be closed manually. Sectionalisers with electronic operating mechanisms are more flexible in operation and easier to set. The load current is measured by means of CTs and the secondary current is fed to a control circuit which counts the number of operations of the recloser or the associated interrupter and then sends a tripping signal to the opening mechanism. This type of sectionaliser is constructed with manual or motor closing.

The following factors should be considered when selecting a sectionaliser:

1. System voltage.
2. Maximum load current.
3. Maximum short-circuit level.
4. Co-ordination with protection devices installed upstream and downstream.

The nominal voltage and current of a sectionaliser should be equal to or greater than the maximum values of voltage or load at the point of installation. The short-circuit capacity (momentary rating) of a sectionaliser should be equal to or greater than the fault level at the point of installation. The maximum clearance time of the associated interrupter should not be permitted to exceed the short-circuit rating of the sectionaliser. Co-ordination factors that need to be taken into account include the starting current setting and the number of operations of the associated interrupter before opening.

6.1.3 Fuses

A fuse is an overcurrent protection device; it possesses an element that is directly heated by the passage of current and is destroyed when the current exceeds a predetermined value. A suitably selected fuse should open the circuit by the destruction of the fuse element, eliminate the arc established during the destruction of the element and then maintain circuit conditions open with nominal voltage applied to its terminals, (i.e. no arcing across the fuse element).

The majority of fuses used in distribution systems operate on the expulsion principle, i.e. they have a tube to confine the arc, with the interior covered with de-ionising fibre, and a fusible element. In the presence of a fault, the interior fibre is heated up when the fusible element melts and produces de-ionising gases which accumulate in the tube. The arc is compressed and expelled out of the tube; in addition, the escape of gas from the ends of the tube causes the particles that sustain the arc to be expelled. In this way, the arc is extinguished when current zero is reached. The presence of de-ionising gases, and the turbulence within the tube, ensure that the fault current is not re-established after the current passes through zero point. The zone of operation is limited by two factors; the lower limit based on the minimum time required for the fusing of the element (minimum melting time) with the upper limit determined by the maximum total time that the fuse takes to clear the fault.

There are a number of standards to classify fuses according to the rated voltages, rated currents, time/current characteristics, manufacturing features and other considerations. For example, there are several sections of ANSI/UL 198-1982 standards that cover low voltage fuses of 600 V or less. For medium and high voltage fuses within the range 2.3–138 kV, standards such as ANSI/IEEE C37.40, 41, 42, 46, 47 and 48 apply. Other organisations and countries have their own standards; in addition, fuse manufacturers have their own classifications and designations.

In distribution systems, the use of fuse links designated K and T for fast and slow types, respectively, depending on the speed ratio, is very popular. The speed ratio is the ratio of minimum melt current that causes fuse operation at 0.1 s to the minimum-melt current for 300 s operation. For the K link, a speed ratio (SR) of 6-8 is defined and, for a T link, 10-13. Figure 6.3 shows the comparative operating characteristics of type 200 K and 200 T fuse links. For the 200 K fuse a 4400 A current is required

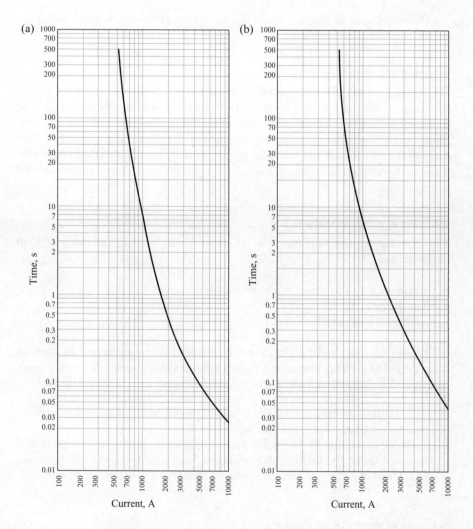

Figure 6.3   *Time/current characteristics of typical fuse links:* (a) 200 K fuse link; (b) 200 T fuse link

for 0.1 s clearance time and 560 A for 300 s, giving an SR of 7.86. For the 200 T fuse, 6500 A is required for 0.1 s clearance, and 520 A for 300 s; for this case, the SR is 12.5.

The following information is required in order to select a suitable fuse for use on the distribution system:

1.   Voltage and insulation level.
2.   Type of system.
3.   Maximum short-circuit level.
4.   Load current.

The above four factors determine the fuse nominal current, voltage and short-circuit capability characteristics.

Selection of nominal current

The nominal current of the fuse should be greater than the maximum continuous load current at which the fuse will operate. An overload percentage should be allowed according to the protected-equipment conditions. In the case of power transformers, fuses should be selected such that the time/current characteristic is above the inrush curve of the transformer and below its thermal limit. Some manufacturers have produced tables to assist in the proper fuse selection for different ratings and connection arrangements.

Selection of nominal voltage

The nominal voltage of the fuse is determined by the following system characteristics:

- maximum phase-to-phase or phase-to-earth voltage;
- type of earthing;
- number of phases (three or one).

The system characteristics determine the voltage seen by the fuse at the moment when the fault current is interrupted. Such a voltage should be equal to, or less than, the nominal voltage of the fuse. Therefore, the following criteria should be used:

1.   In unearthed systems, the nominal voltage should be equal to, or greater than, the maximum phase-to-phase voltage.
2.   In three-phase earthed systems, for single-phase loads the nominal voltage should be equal to, or greater than, the maximum line-to-earth voltage and for three-phase loads the nominal voltage is selected on the basis of the line-to-line voltage.

Selection of short-circuit capacity

The symmetrical short-circuit capacity of the fuse should be equal to, or greater than, the symmetrical fault current calculated for the point of installation of the fuse.

Fuse notation

When two or more fuses are used on a system, the device nearest to the load is called the main protection, and that upstream, towards the source, is called the back up. The criteria for co-ordinating them will be discussed later.

6.2 Criteria for co-ordination of time/current devices in distribution systems

The following basic criteria should be employed when co-ordinating time/current devices in distribution systems:

1. The main protection should clear a permanent or temporary fault before the back-up protection operates, or continue to operate until the circuit is disconnected. However, if the main protection is a fuse and the back-up protection is a recloser, it is normally acceptable to co-ordinate the fast operating curve or curves of the recloser to operate first, followed by the fuse, if the fault is not cleared. (See Section 6.2.2.)
2. Loss of supply caused by permanent faults should be restricted to the smallest part of the system for the shortest time possible.

In the following sections criteria and recommendations are given for the co-ordination of different devices used on distribution systems.

6.2.1 Fuse-fuse co-ordination

The essential criterion when using fuses is that the maximum clearance time for a main fuse should not exceed 75 per cent of the minimum melting time of the back-up fuse, for the same current level, as indicated in Figure 6.4. This ensures that the main fuse interrupts and clears the fault before the back-up fuse is affected in any way. The factor of 75 per cent compensates for effects such as load current and ambient temperature, or fatigue in the fuse element caused by the heating effect of fault currents that have passed through the fuse to a fault downstream but were not sufficiently large enough to melt the fuse.

The co-ordination between two or more consecutive fuses can be achieved by drawing their time/current characteristics, normally on log-log paper as for over-current relays. In the past, co-ordination tables with data of the available fuses were also used, which proved to be an easy and accurate method. However, the graphic method is still popular not only because it gives more information but also because computer-assisted design tools make it much easier to draw out the various characteristics.

6.2.2 Recloser-fuse co-ordination

The criteria for determining recloser-fuse co-ordination depend on the relative locations of these devices, i.e. whether the fuse is at the source side and then backs up the operation of the recloser that is at the load side, or vice versa. These possibilities are treated in the following paragraphs.

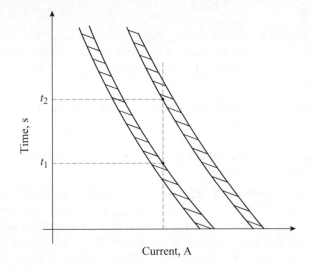

Figure 6.4   Criteria for fuse-fuse co-ordination: $t_1 < 0.75\, t_2$

Fuse at the source side

When the fuse is at the source side, all the recloser operations should be faster than the minimum melting time of the fuse. This can be achieved through the use of multiplying factors on the recloser time/current curve to compensate for the fatigue of the fuse link produced by the cumulative heating effect generated by successive recloser operations. The recloser opening curve modified by the appropriate factor then becomes slower but, even so, should be faster than the fuse curve. This is illustrated in Figure 6.5.

The multiplying factors referred to above depend on the reclosing time in cycles and on the number of the reclosing attempts. Some values proposed by Cooper Power Systems are reproduced in Table 6.1.

It is convenient to mention that if the fuse is at the high voltage side of a power transformer and the recloser at the low voltage side, either the fuse or the recloser curve should be shifted horizontally on the current axis to allow for the transformer turns ratio. Normally it is easier to shift the fuse curve, based on the transformer tap that produces the highest current on the high voltage side. On the other hand, if the transformer connection group is delta-star, the considerations given in Section 5.5 should be taken into account.

Fuses at the load side

The procedure to co-ordinate a recloser and a fuse, when the latter is at the load side, is carried out with the following rules:

- the minimum melting time of the fuse must be greater than the fast curve of the recloser times the multiplying factor, given in Table 6.2 and taken from the same reference as above;

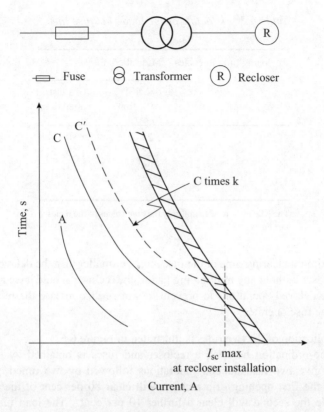

Figure 6.5   Criteria for source-side fuse and recloser co-ordination

Table 6.1   k factor for source-side fuse link

Reclosing time in cycles	Multipliers for:		
	two fast, two delayed sequence	one fast, three delayed sequence	four delayed sequence
25	2.70	3.20	3.70
30	2.60	3.10	3.50
50	2.10	2.50	2.70
90	1.85	2.10	2.20
120	1.70	1.80	1.90
240	1.40	1.40	1.45
600	1.35	1.35	1.35

The *k* factor is used to multiply the time values of the delayed curve of the recloser.

Table 6.2   k factor for the load-side fuse link

Reclosing time in cycles	Multipliers for:	
	one fast operation	two fast operations
25–30	1.25	1.80
60	1.25	1.35
90	1.25	1.35
120	1.25	1.35

The *k* factor is used to multiply the time values of the recloser fast curve.

- the maximum clearing time of the fuse must be smaller than the delayed curve of the recloser without any multiplying factor; the recloser should have at least two or more delayed operations to prevent loss of service in case the recloser trips when the fuse operates.

The application of the two rules is illustrated in Figure 6.6.

Better co-ordination between a recloser and fuses is obtained by setting the recloser to give two instantaneous operations followed by two timed operations. In general, the first opening of a recloser will clear 80 per cent of the temporary faults, while the second will clear a further 10 per cent . The load fuses are set to operate before the third opening, clearing permanent faults. A less effective co-ordination is obtained using one instantaneous operation followed by three timed operations.

6.2.3   Recloser-recloser co-ordination

The co-ordination between reclosers is obtained by appropriately selecting the amperes setting of the trip coil in the hydraulic reclosers, or of the pick-ups in electronic reclosers.

Hydraulic reclosers

The co-ordination margins with hydraulic reclosers depend upon the type of equipment used. In small reclosers, where the current coil and its piston produce the opening of the contacts, the following criteria must be taken into account:

- separation of the curves by less than two cycles always results in simultaneous operation;
- separation of the curves by between two and 12 cycles could result in simultaneous operation;
- separation greater than 12 cycles ensures nonsimultaneous operation.

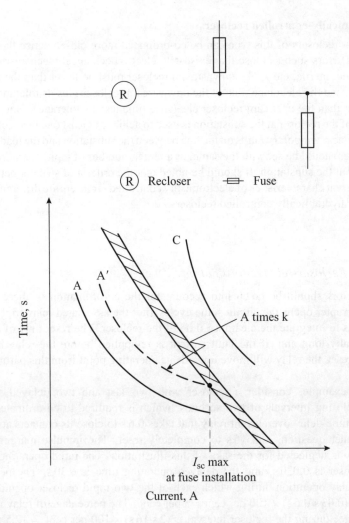

Figure 6.6   Criteria for recloser and load-side fuse co-ordination

With large capacity reclosers, the piston associated with the current coil only actuates the opening mechanism. In such cases, the co-ordination margins are as follows:

- separation of the curves by less than two cycles always results in simultaneous operation;
- a separation of more than eight cycles guarantees non-simultaneous operation.

The principle of co-ordination between two large units in series is based on the time of separation between the operating characteristics, in the same way as for small units.

Electronically-controlled reclosers

Adjacent reclosers of this type can be co-ordinated more closely since there are no inherent errors such as those that exist with electromechanical mechanisms (due to overspeed, inertia, etc.). The downstream recloser must be faster than the upstream recloser, and the clearance time of the downstream recloser plus its tolerance should be lower than the upstream recloser clearance time less its tolerance. Normally, the setting of the recloser at the substation is used to achieve at least one fast reclosure, in order to clear temporary faults on the line between the substation and the load recloser. The latter should be set with the same, or a larger, number of rapid operations as the recloser at the substation. It should be noted that the criteria of spacing between the time/current characteristics of electronically controlled reclosers are different to those used for hydraulically controlled reclosers.

6.2.4   Recloser-relay co-ordination

Two factors should be taken into account for the co-ordination of these devices; the interrupter opens the circuit some cycles after the associated relay trips, and the relay has to integrate the clearance time of the recloser. The reset time of the relay is normally long and, if the fault current is re-applied before the relay has completely reset, the relay will move towards its operating point from this partially reset position.

For example, consider a recloser with two fast and two delayed sequence with reclosing intervals of two seconds, which is required to co-ordinate with an inverse time-delay overcurrent relay that takes 0.6 s to close its contacts at the fault level under question, and 16 s to completely reset. The impulse margin time of the relay is neglected for the sake of this illustration. The rapid operating time of the recloser is 0.030 s, and the delayed operating time is 0.30 s. The percentage of the relay operation during which each of the two rapid recloser openings takes place is $(0.03 \text{ s}/0.6 \text{ s}) \times 100$ per cent $= 5$ per cent . The percentage of relay reset that takes place during the recloser interval is $(2 \text{ s}/16 \text{ s}) \times 100$ per cent $= 12.5$ per cent. Therefore, the relay completely resets after both of the two rapid openings of the recloser.

The percentage of the relay operation during the first time-delay opening of the recloser is $(0.3 \text{ s}/0.6 \text{ s}) \times 100$ per cent $= 50$ per cent. The relay reset for the third opening of the recloser $= 12.5$ per cent, as previously, so that the net percentage of relay operation after the third opening of the recloser $= 50$ per cent $- 12.5$ per cent $= 37.5$ per cent . The percentage of the relay operation during the second time delay opening of the recloser takes place $= (0.3 \text{ sec.}/0.6 \text{ sec}) \times 100$ per cent $= 50$ per cent , and the total percentage of the relay operation after the fourth opening of the recloser $= 37.5$ per cent $+ 50$ per cent $= 87.5$ per cent.

From the above analysis it can be concluded that the relay does not reach 100 per cent operation by the time the final opening shot starts, and therefore co-ordination is guaranteed.

6.2.5 Recloser-sectionaliser co-ordination

Since the sectionalisers have no time/current operating characteristic, their co-ordination does not require an analysis of these curves.

The co-ordination criteria in this case are based upon the number of operations of the back-up recloser. These operations can be any combination of rapid or timed shots as mentioned previously, for example two fast and two delayed. The sectionaliser should be set for one shot less than those of the recloser, for example three disconnections in this case. If a permanent fault occurs beyond the sectionaliser, the sectionaliser will open and isolate the fault after the third opening of the recloser. The recloser will then re-energise the section to restore the circuit. If additional sectionalisers are installed in series, the furthest recloser should be adjusted for a smaller number of counts. A fault beyond the last sectionaliser results in the operation of the recloser and the start of the counters in all the sectionalisers.

6.2.6 Recloser-sectionaliser-fuse co-ordination

Each one of the devices should be adjusted in order to co-ordinate with the recloser. In turn, the sequence of operation of the recloser should be adjusted in order to obtain the appropriate co-ordination for faults beyond the fuse by following the criteria already mentioned.

Example 6.1

Figure 6.7 shows a portion of a 13.2 kV distribution feeder that is protected by a set of overcurrent relays at the substation location. A recloser and a sectionaliser have been installed downstream to improve the reliability of supply to customers. The recloser chosen has two fast and two delayed operations with 90 cycles intervals.

The time/current curves for the transformer and branch fuses, the recloser and the relays, are shown in Figure 6.8. For a fault at the distribution transformer, its fuse should operate first, being backed up by the recloser fast operating shots. If the fault is still not cleared, then the branch fuse should operate next followed by the delayed opening shots of the recloser and finally by the operation of the feeder relay. The sectionaliser will isolate the faulted section of the network after the full number of counts has elapsed, leaving that part of the feeder upstream still in service.

As the nominal current of the 112.5 kVA distribution transformer at 13.2 kV is 4.9 A, a 6 T fuse was selected on the basis of allowing a 20 per cent overload. The fast curve of the recloser was chosen with the help of the following expression based on the criteria already given, which guarantees that it lies in between the curves of both fuses:

$$t_{\text{recloser}} \times k \leq t_{\text{MMT of branch fuse}} \times 0.75 \tag{6.1}$$

where $t_{\text{MMT of branch fuse}}$ is the minimum melting time. The 0.75 factor is used in order to guarantee the co-ordination of the branch and transformer fuses, as indicated in Section 6.2.1.

At the branch fuse location the short-circuit current is 2224 A, which results in operation of the branch fuse in 0.02 s. From Table 6.2, the k factor for two fast

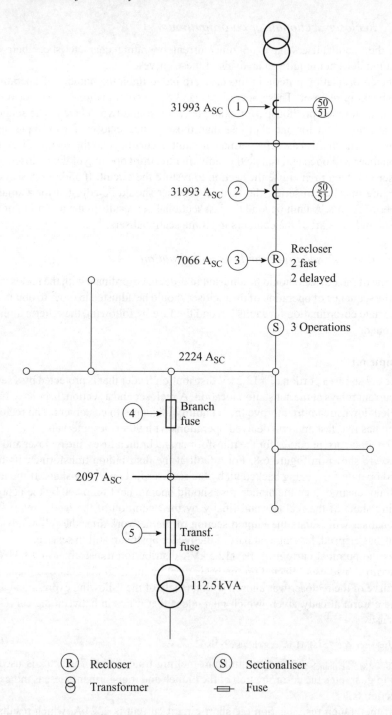

Figure 6.7   Portion of a distribution feeder for Example 6.1

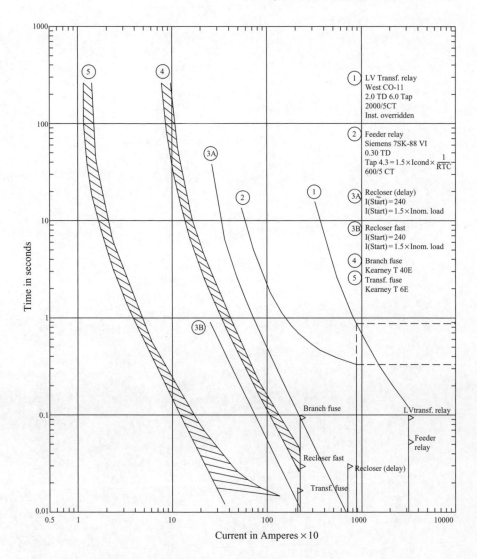

Figure 6.8   Phase-current curves for Example 6.1

operations and a reclosing time of 90 cycles is 1.35. With these values, from eqn. 6.1 the maximum time for the recloser operation is $(0.02 \times 0.75/1.35) = 0.011$ s. This time, and the pick-up current of the recloser, determines the fast curve of the recloser.

The feeder relay curve is selected so that it is above that of the delayed curve of the recloser, and so that the relay reset time is considered. Detailed calculations have not been given for this particular example since the procedure was indicated in Chapter 5, but the curves of Figure 6.8 shows that adequate co-ordination has been achieved.

Chapter 7

Directional overcurrent relays

Directional overcurrent protection is used when it is necessary to protect the system against fault currents that could circulate in both directions through a system element, and when bi-directional overcurrent protection could produce unnecessary disconnection of circuits. This can happen in ring or mesh-type systems and in systems with a number of infeed points. The use of directional overcurrent relays in the two situations is shown in Figure 7.1.

7.1 Construction

Directional overcurrent relays are constructed using a normal overcurrent unit plus a unit that can determine the direction of the power flow in the associated distribution system element. In addition to the relay current this second unit usually requires a reference signal to measure the angle of the fault and thus determine whether or not

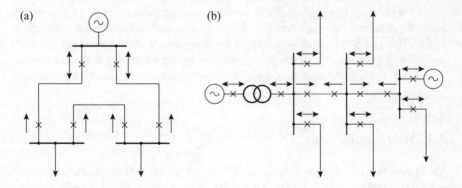

Figure 7.1   Application of directional overcurrent relays: (a) ring system; (b) multi-source system

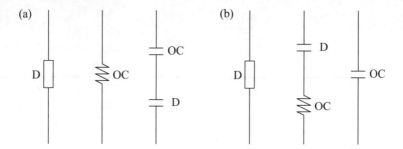

Figure 7.2   Obtaining the direction of power flow: (a) by supervision; (b) by control

the relay should operate. Generally, the reference or polarisation signal is a voltage but this can also be a current input.

Basically, there are two methods for obtaining the direction of the power flow – supervision and control; both cases are illustrated in Figure 7.2 where D indicates the directional unit and OC the overcurrent unit. It is better to use the control system to determine the direction of the power flow, since the overcurrent unit only picks up when the flow is in the correct direction. With the supervision method, the overcurrent unit can pick up for the wrong power flow direction. In addition, when a breaker is opened in a ring system the current flows will change and this could lead to the consequential possibility of loss of co-ordination.

7.2   Principle of operation

The operating torque can be defined by $T = K\Phi_1\Phi_2\sin\Theta$ where Φ_1 and Φ_2 are the polarising values, Φ_1 being proportional to the current and Φ_2 proportional to the voltage, with Θ the angle between Φ_1 and Φ_2. The torque is positive if $0 < \Theta < 180°$ and negative if $180° < \Theta < 360°$. It should be noted that Θ is in phase with I but lagging with respect to the voltage since $V = -(d\Phi)/dt$.

If I and V are in phase, then the fluxes are out of phase by $90°$. Therefore, the angle for maximum torque occurs when the current and voltage of the relay are in phase. This can be obtained very simply by using the current and voltage from the same phase. However, this is not practical since, for a fault on one phase, the voltage of that phase might collapse. It is, therefore, common practice to use the current from a different phase.

7.3   Relay connections

The connection of a directional relay is defined on the basis of the degrees by which a current at unity power factor leads the polarisation voltage. The angle of maximum torque, AMT, is the angle for which this displacement produces the maximum torque and therefore is always aligned with the polarisation voltage.

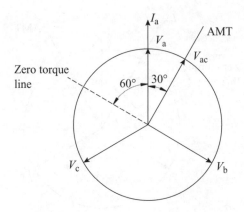

Figure 7.3   Vector diagram for 30° connection (0° AMT)

7.3.1   30° connection (0° AMT)

Feeding the relays:

$$\Phi_A: \quad I_a, \quad \Phi_B: \quad I_b, \quad \Phi_C: \quad I_c$$
$$V_{ac} \qquad\qquad V_{ba} \qquad\qquad V_{cb}$$

Maximum torque: when the phase current lags the phase-neutral voltage by 30°.
Angle of operation: current angles from 60° leading to 120° lagging.
Use: this type of connection should always be used on feeders, provided that it has three elements, i.e. one per phase, since two phase elements and one earth element can give rise to poor operation (see Figure 7.3). The three-phase unit arrangement should not be used in transformer circuits where some faults can result in a reverse current flow in one or more phases leading to relay mal-operation.

7.3.2   60° connection (0° AMT)

Feeding the relays:

$$\Phi_A: \quad I_{ab}, \quad \Phi_B: \quad I_{bc}, \quad \Phi_C: \quad I_{ca}$$
$$V_{ac} \qquad\qquad V_{ba} \qquad\qquad V_{cb}$$

Maximum torque: when the current lags the phase-to-neutral voltage by 60°. I_{ab} lags V_{ac} by 60°. I_a lags V_a by 60° at unity power factor.
Angle of operation: current I_{ab} from 30° leading to 150° lagging, or I_a leading 30° or lagging 150° at unity power factor.
Use: it is recommended that relays with this connection be used solely on feeders (see Figure 7.4). However, they have the disadvantage that the CTs have to be connected in delta. For this reason, and because they offer no advantages compared to the previous case, they are little used.

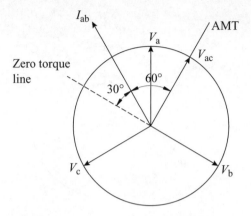

Figure 7.4   Vector diagram for 60° connection (0° AMT)

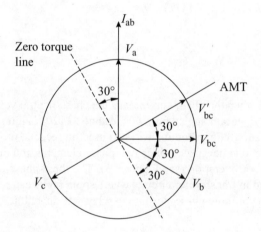

Figure 7.5   Vector diagram for 90° connection (30° AMT)

7.3.3   90° connection (30° AMT)

Feeding the relays:

$$\Phi_A: \quad I_a, \qquad \Phi_B: \quad I_b; \qquad \Phi_C: \quad I_c$$
$$\qquad\quad V_{bc} + 30° \qquad\qquad V_{ca} + 30° \qquad\qquad V_{ab} + 30°$$

Maximum torque: when the current lags the phase-to-neutral voltage by 60°.
Angle of operation: current angles from 30° leading to 150° lagging.
Use: in feeders where the source of zero-sequence components is behind the point of connection of the relay (see Figure 7.5).

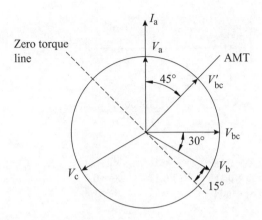

Figure 7.6   Vector diagram for 90° connection (45° AMT)

7.3.4   *90° connection (45° AMT)*

Feeding the relays:

$$\Phi_A: \quad I_a, \qquad \Phi_B: \quad I_b, \qquad \Phi_C: \quad I_c,$$
$$V_{bc} + 45° \qquad\qquad V_{ca} + 45° \qquad\qquad V_{ab} + 45°$$

Maximum torque: when the current lags the phase-to-neutral voltage by 45°.

Angle of operation: current angles from 45° leading to 135° lagging.

Use: this arrangement is recommended for protecting transformers or feeders that have a source of zero-sequence components in front of the relay (see Figure 7.6). This connection is essential in the case of transformers in parallel, or transformer feeders, especially for guaranteeing correct operation of the relays for faults beyond Yd transformers. This configuration should always be used when single-phase directional relays are applied to circuits that can have a 2-1-1 current distribution.

7.4   Directional earth-fault relays

Directional earth-fault relays are constructed on the basis that the residual voltage is equal to three times the zero-sequence voltage drop in the source impedance, and displaced with respect to the residual current for the angle characteristic of the source impedance. When a set of suitable VTs is not available for obtaining the polarisation voltage, current polarisation is employed using the earth current from a local transformer connected to earth. This is based on the principle that the neutral current always flows towards the system from earth whereas, depending on the fault, the residual current may flow in any direction. It should be stressed, however, that the possibility of the failure of a voltage-polarised directional protection relay is minimal and it is therefore recommended that this arrangement should be used wherever possible.

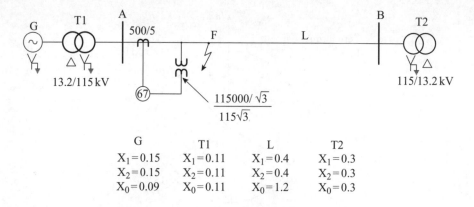

Figure 7.7   Single line diagram of system for Example 7.1

Example 7.1

A solid earth-fault on phases B and C is represented by the arrow at the point F in the power system in Figure 7.7. Determine the current and voltage signals (in amps and volts) that go into each one of the directional relays that have a 30° connection and are fed as indicated below:

$$\Phi_A: \quad I_A, \quad \Phi_B: \quad I_B, \quad \Phi_C: \quad I_C$$
$$\qquad V_{AC} \qquad\quad V_{BA} \qquad\quad V_{CB}$$

In addition, indicate which relays operate on the occurrence of the fault. In the solution, ignore load currents and assume a pre-fault voltage equal to 1 p.u. The bases at the generator location are 13.2 kV and 100 MVA.

N.B. Although the system is radial, the installation of a directional overcurrent relay is justified by assuming that this circuit would be part of a ring in a future system.

Solution

The conditions for a double phase-to-earth fault, B-C-N, are

$$I_A = 0, V_B = 0, V_C = 0$$

The three sequence networks are shown in Figure 7.8. The equivalent circuit is obtained by connecting the three sequence networks in parallel as shown in Figure 7.9.
   From Figure 7.8

$$Z_0 = 0.11 \parallel 1.5 \Rightarrow Z_0 = 0.102$$

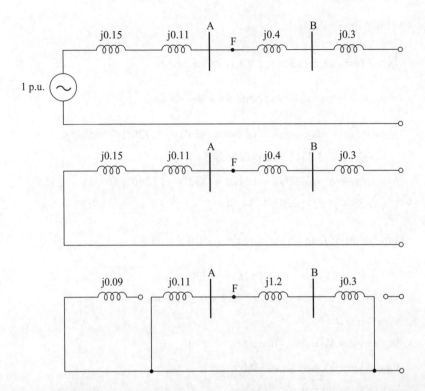

Figure 7.8   Sequence networks for Example 7.1

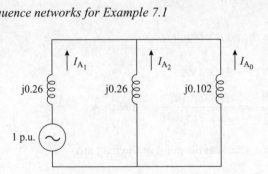

Figure 7.9   Equivalent circuit for Figure 7.8

so that the three sequence currents in the network are

$$I_{A1} = \frac{1}{(j0.26 + (j0.26 \times j0.102)/(j0.362))} = -j3.0 \text{p.u.}$$

$$I_{A2} = -(-j3.0)\frac{j0.102}{j0.26 + j0.102} = j0.845$$

$$I_{A0} = -(-j3.0)\frac{j0.26}{j0.26 + j0.102} = j2.155$$

At the point of the fault:

$$I_A = I_{A1} + I_{A2} + I_{A0} = -j3.0 + j0.845 + j2.155$$

$$\Rightarrow I_A = 0 \text{ as was to be expected for a B-C-N fault}$$

$$I_B = a^2 I_{A1} + a I_{A2} + I_{A0} = 1\angle240°(-j3.0) + 1\angle120°(j0.845) + j2.155$$
$$= -3.33 + j3.2315 = 4.64\angle135.86°$$
$$I_C = a I_{A1} + a^2 I_{A2} + I_{A0} = 1\angle120°(-j3.0) + 1\angle240°(j0.845) + j2.155$$
$$= 3.33 + j3.2315 = 4.64\angle44.14°$$

At the point of fault on the network (not at the relay):

$$V_{A1} = V_{A2} = V_{A0} = -I_{A2}(j0.26) = 0.220$$
$$V_A = V_{A1} + V_{A2} + V_{A0} = 3V_{A1} = 3 \times 0.22 = 0.66$$
$$V_B = V_C = 0$$
$$V_{AC} = V_A - V_C = V_A - 0 = 0.66$$
$$V_{BA} = V_B - V_A = -V_A = -0.66$$
$$V_{CB} = V_C - V_B = 0$$

The bases at the point of fault are

$$V = 115\,\text{kV}; \ P = 100\,\text{MVA}$$

$$I_{Base} = \frac{P}{\sqrt{3}V} = \frac{100 \times 10^6}{\sqrt{3} \times 115 \times 10^3} = 502.04\,\text{A}$$

Therefore, the values at the point of the fault are

$$I_A = 0$$

$$V_{AC} = 0.66 \times \frac{115000}{\sqrt{3}} \times \frac{115}{\sqrt{3}} \times \frac{\sqrt{3}}{115000} = 43.82\angle0° \text{ V}$$

$$I_B = 4.64\angle135.86° \times 502.04 \times (5/500) = 23.29\angle135.86° \text{ A}$$

$$V_{BA} = -0.66 \times \frac{115000}{\sqrt{3}} \times \frac{1}{1000} = 43.82\angle180° \text{ V}$$

$$I_C = 4.64\angle44.14° \times 502.04 \times (5/500) = 23.29\angle44.14° \text{ A}$$

$$V_{CB} = 0$$

Corollary

$$I_{3\phi} = \frac{1}{j0.26} = -j3.48 \quad I_{1\phi} = I_{A1} + I_{A2} + I_{A0} = 3I_{A1}$$

and

$$I_{A1} = \frac{1}{j0.26 + j0.26 + j0.102} = -j1.6077$$

so that

$$I_{1\phi} = 3 \times (-j1.6077) = -j4.823$$

At the point where the relay is located there will be equal positive- and negative-sequence current values, as there are at the fault point; however, the zero-sequence current in the relay itself is different because of the division of current.

In the relay, $I_{AB} = j2.154 (1.5/1.61) = j2$ p.u., so that at the relay

$$I_A = I_{A1} + I_{A2} + I_{A0} = -j3.0 + j0.845 + j2 = -j0.155$$

and it will be seen that, in this case, $I_A \neq 0$:

$$I_B = a^2 I_{A1} + a I_{A2} + I_{A0} = 1\angle 240°(-j3.0) + 1\angle 120°(j0.845) + j2$$
$$= 3.0\angle 150° + 0.845\angle 210° + 2.0\angle 90°$$
$$= -3.33 + j3.077 = 4.534\angle 137.26°$$

$$I_C = a I_{A1} + a^2 I_{A2} + I_{A0} = 1\angle 120°(-j3.0) + 1\angle 240°(j0.845) + j2$$
$$= 3.0\angle 30° + 0.845\angle -30° + 2.0\angle 90°$$
$$= 3.33 + j3.077 = 4.534\angle 42.73°$$

$$V_{A1} = V_{A2} = I_{A2}(j0.26) = 0.845 \times 0.26 = 0.22$$
$$V_{A0} = -I_{A0}(j0.11) = -j2.0 (j0.11) = 0.22$$
$$V_A = V_{A1} + V_{A2} + V_{A0} = 3V_{A1} = 3 \times 0.22 = 0.66$$
$$V_B = V_C = 0$$

The CT is fed from the same fault point, so that

$$V_{AC} = V_A - V_C = V_A - 0 = 0.66$$
$$V_{BA} = V_B - V_A = -V_A = -0.66$$
$$V_{CB} = V_C - V_B = 0$$

The signals that feed the relay are

Φ_A

$$I_A = 0.155\angle-90° \times 502.04 \times (5/500) = 0.728\angle-90° \text{ A}$$

$$V_{AC} = 0.66 \times \frac{115000}{\sqrt{3}} \times \frac{115\sqrt{3}}{115000/\sqrt{3}} = 43.82\angle0° \text{ V}$$

Φ_B

$$I_B = 4.534\angle137.26° \times 502.04 \times (5/500) = 22.76\angle137.26° \text{ A}$$

$$V_{BA} = -0.66 \times \frac{115000}{\sqrt{3}} \times \frac{1}{1000} = 43.82\angle180° \text{ V}$$

Φ_C

$$I_C = 4.534\angle42.73° \times 502.04 \times (5/500) = 22.76\angle42.73° \text{ A}$$

$$V_{BC} = 0$$

Analysis of operation of directional relays:

Polarisation:

Φ_A	Φ_B	Φ_C
I_A	I_B	I_C
V_{AC}	V_{BA}	V_{CB}

Phase A relay:

$$I_A = 0.728\angle-90° \text{ A}$$

$$V_{AC} = 43.82\angle-0° \text{ V}$$

For operation, $-90° <$ angle of $I_A < 90°$. The relay in phase A is at the limit of functioning (see Figure 7.10) thus creating some doubt about the operation of its directional unit.

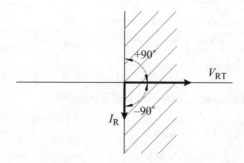

Figure 7.10   Analysis of operation of relay for phase A

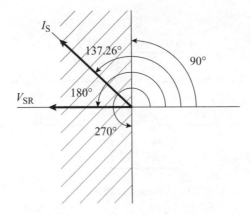

Figure 7.11   Analysis of operation of relay for phase B

Phase B relay:

$$I_B = 22.76 \angle 137.26° \text{ A}$$

$$V_{BA} = 43.82 \angle 180° \text{ V}$$

For operation, $90° <$ angle of $I_B < 270°$. The relay in phase B operates, since the angle of I_B is $137.26°$; see Figure 7.11.

Phase C relay: this does not operate because $V_{CB} = 0$.

7.5   Co-ordination of instantaneous units

The calculations for the setting of an instantaneous unit in a ring system are carried out using the short-circuit level at the next relay downstream, with the ring open, multiplied by a safety overload factor in order to maintain co-ordination, taking into account the DC transient component of the current. The criterion used here is the same as for setting bi-directional overcurrent relays protecting lines between substations. When the ring has only one source, the relays installed at substations adjacent to the source substation should never register any current from the substation towards the source. Therefore, it is recommended that, for these relays, the instantaneous units should be set at 1.5 times the maximum load current. A lower value should not be used as this could result in a false trip if the directional element picks up inadvertently under severe load transfer conditions. With these settings the instantaneous units then have the same pick-up current value as that of the time-delay units so that co-ordination is not compromised.

Example 7.2

For the system shown in Figure 7.12 determine the maximum load currents, the transformation ratios of the CTs and the current setting of the instantaneous units

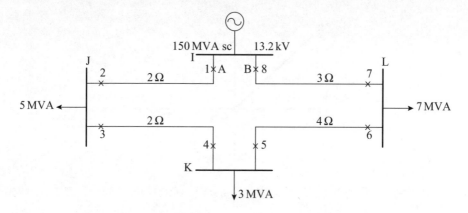

Figure 7.12   Single-line diagram for a ring system with one infeed for Example 7.2

to guarantee a co-ordinated protection scheme. The instantaneous settings should be given in primary amperes, since reference to specific relay models is not given.

Calculation of the maximum load currents

With the ring open at A:

		Relays
total load current flowing through B	$= (7+3+5 \text{ MVA})/\sqrt{3}(13200 \text{ V})$	
	$= (15 \text{ MVA})/\sqrt{3}(13200 \text{ V}) = 656.08 \text{ A}$	(7) (8)
total load current from L to K	$= (8 \text{ MVA})/\sqrt{3}(13200 \text{ V}) = 349.9 \text{ A}$	(5) (6)
total load current from K to J	$= (5 \text{ MVA})/\sqrt{3}(13200 \text{ V}) = 218.69 \text{ A}$	(3) (4)
total load current flowing from J	$= 0 \text{ A}$	(1) (2)

With the ring open at B:

total load current flowing through A	$= (15 \text{ MVA})/\sqrt{3}(13200 \text{ V}) = 656.08 \text{ A}$	(1) (2)
total load current from J to K	$= (10 \text{ MVA})/\sqrt{3}(13200 \text{ V}) = 437.38 \text{ A}$	(3) (4)
total load current from K to L	$= (7 \text{ MVA})/\sqrt{3}(13200 \text{ V}) = 306.17 \text{ A}$	(5) (6)
total load flowing from L	$= 0 \text{ A}$	(7) (8)

Selection of CTs

It is assumed that the available CTs have primary turns in multiples of 100 up to 600, and thereafter in multiples of 200. The CT ratios are calculated for the maximum load conditions first. Thus, the CT ratios selected are:

$I_{\text{max} 1} = 656.08 \text{ A}$ CT ratio $= 800/5$

$I_{\text{max} 2} = 656.08 \text{ A}$ CT ratio $= 800/5$

$I_{\text{max} 3} = 437.38 \text{ A}$ CT ratio $= 500/5$

$I_{\text{max} 4} = 437.38 \text{ A}$ CT ratio $= 500/5$

$I_{\text{max} 5} = 349.90 \text{ A}$ CT ratio $= 400/5$

$I_{\max 6} = 349.90\,\mathrm{A}$ CT ratio $= 400/5$

$I_{\max 7} = 656.08\,\mathrm{A}$ CT ratio $= 800/5$

$I_{\max 8} = 656.08\,\mathrm{A}$ CT ratio $= 800/5$

In order to confirm this selection, it is necessary to check whether or not saturation is present at the maximum fault level at each breaker, using the above CT ratios. With the most critical fault values, as calculated in the following section, it can be shown that no CT is saturated since the value of $0.05 \times I_{sc}$ is well below the number of primary turns for each one.

Calculation of fault currents
Busbar I

$$I_{sc} = (150\,\mathrm{MVA})/\sqrt{3}(13200\,\mathrm{V}) = 6560.8\,\mathrm{A}$$

$$Z_{\text{source}} = (13200\,\mathrm{V})^2/(150\,\mathrm{MVA}) = 1.16\,\Omega$$

Busbar J

The equivalent circuit for a fault at J is shown in Figure 7.13:

$$I_{scJ} = \frac{13200}{\sqrt{3}(1.16 + 1.64)} = 2721.79\,\mathrm{A}, \text{ with the ring closed.}$$

Apportioning the current flows in the inverse proportion of the circuit impedances:

I_{scJ} (on the right hand side) $= 2721.79 \times (2 \div 11) = 494.87\,\mathrm{A}$

I_{scJ} (on the left hand side) $= 2721.79 \times (9 \div 11) = 2226.92\,\mathrm{A}$

With breaker A open:

$$I_{scJ} = \frac{13200}{\sqrt{3}(1.16 + 9)} = 750.1\,\mathrm{A}$$

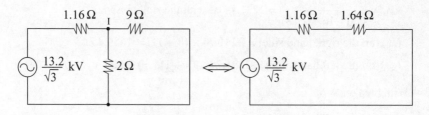

Figure 7.13   Equivalent circuit for a fault at J (Example 7.2)

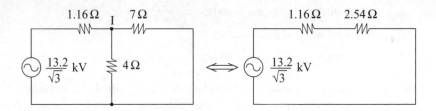

Figure 7.14   Equivalent circuit for a fault at K (Example 7.2)

With breaker B open:

$$I_{scJ} = \frac{13200}{\sqrt{3}(1.16+2)} = 2411.71\,\text{A}$$

Busbar K

The equivalent circuit for a fault at K is shown in Figure 7.14:

$$I_{scK} = \frac{13200}{\sqrt{3}(1.16+2.54)} = 2059.74\,\text{A, with the ring closed.}$$

I_{scK} (on the right hand side) $= 2059.74 \times (4 \div 11) = 748.99\,\text{A}$

I_{scK} (on the left hand side) $= 2059.74 \times (4 \div 11) = 1310.74\,\text{A}$

With breaker A open:

$$I_{scK} = \frac{13200}{\sqrt{3}(1.16+7)} = 933.94\,\text{A}$$

With breaker B open:

$$I_{scK} = \frac{13200}{\sqrt{3}(1.16+4)} = 1476.94\,\text{A}$$

Busbar L

The equivalent circuit for a fault at L is shown in Figure 7.15:

$$I_{scL} = \frac{13200}{\sqrt{3}(1.16+2.18)} = 2281.74\,\text{A, with the ring closed.}$$

I_{scL} (on the right hand side) $= 2281.74 \times (8 \div 11) = 1659.45\,\text{A}$

I_{scL} (on the left hand side) $= 2281.74 \times (3 \div 11) = 622.29\,\text{A}$

With breaker A open:

$$I_{scL} = \frac{13200}{\sqrt{3}(1.16+3)} = 1831.97\,\text{A}$$

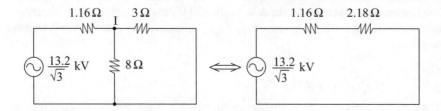

Figure 7.15   Equivalent circuit for a fault at L (Example 7.2)

With breaker B open:

$$I_{scL} = \frac{13200}{\sqrt{3}(1.16+8)} = 831.99 \, \text{A}$$

Setting of instantaneous unit

Clockwise direction:

Relay 7: $1.50 \times 656.08 \, \text{A} = 984.12 \, \text{A}$

Relay 5: $1.25 \times 831.99 \, \text{A} = 1039.99 \, \text{A}$

Relay 3: $1.25 \times 1476.94 \, \text{A} = 1846.17 \, \text{A}$

Relay 1: $1.25 \times 2411.71 \, \text{A} = 3014.64 \, \text{A}$

Anticlockwise direction:

Relay 2: $1.50 \times 656.08 \, \text{A} = 984.12 \, \text{A}$

Relay 4: $1.25 \times 750.10 \, \text{A} = 937.62 \, \text{A}$

Relay 6: $1.25 \times 933.94 \, \text{A} = 1167.42 \, \text{A}$

Relay 8: $1.25 \times 1831.97 \, \text{A} = 2289.96 \, \text{A}$

7.6   Setting of time-delay directional overcurrent units

As in the case of bi-directional overcurrent relays, the time-delay units of directional overcurrent relays in a ring are set by selecting suitable values of the pick-up current and time dial settings. The procedure for each one is indicated in the following sections.

7.6.1   Pick-up setting

The pick-up setting of a directional overcurrent relay is calculated by considering the maximum load transfer that can be seen by the relay in any direction, multiplied by the overload factor referred to in Section 5.3.3. The load transfer in both directions is taken into account to avoid the possibility of relay mal-operation if the directional unit is incorrectly activated by the wrong polarisation, especially under heavy transfer conditions.

7.6.2   Time dial setting

The time dial setting can be defined by means of two procedures. The first one is based on instantaneous setting values, whereas the second takes account of contact travel and is more rigorous since it requires fault calculations for various ring topologies. However, it has to be emphasised that both methods guarantee proper co-ordination although the first one can produce slightly higher time dial values and is more used in simple systems or when the locations of the co-ordination curves are not critical.

Time dial setting by direct method

The setting of time dial units by the direct method is based on the fault values used to set the instantaneous units. As in the case of bi-directional relays, the time dial value is adjusted so that, taking the instantaneous current setting given to the relay downstream, its operating time is above that of the downstream relay by the required discrimination time margin. This procedure should be carried out for all the relays on the ring, both clockwise and anticlockwise, normally starting from the relays associated with the main source busbar. The application of this method is illustrated in the instantaneous settings calculated for the relays in Example 7.2.

Time dial setting considering contact travel

The time dial setting of directional relays, taking account of the contact travel of the timer units, requires an iterative process as detailed below:

1. Determine the initial time dial values of the relays on the ring such that co-ordination is guaranteed with the relays associated with the lines and machines fed by the adjacent busbar in the direction of trip.
2. Calculate the time required for the first relay to operate for a fault at its associated breaker terminals, with the ring closed. Any relay can be chosen as the first, although it is usual to take one of the relays associated with equipment connected to the main source busbar. For this condition a check should be made to ensure that there is adequate discrimination between the chosen relay and the back-up relays at adjacent substations. If not, then the time dial values of the relays at the adjacent substations should be modified. In addition, the operating time of the relay on the breaker at the opposite end of the line should be calculated as well as the times for its back-up relays.
3. Next, consider a fault at the opposite end of the line with the ring open and, for this condition, calculate the operating time of the relay closest to the fault and check that there is adequate discrimination between it and the back-up relays at the adjacent substations. As in the previous case, if co-ordination is not achieved then the time dial values should be increased. For this case it is important to take

into account the contact travel during the fault before the ring is opened by the operation of the first relay. To do this, the following expressions should be used:

$$t_{\text{relay next to fault}} = t_{\text{adjacent relay with ring closed}}$$

$$+ t_{\text{relay next to fault with ring open}} \left(1 - \frac{t_{\text{adjacent relay with ring closed}}}{t_{\text{relay next to fault with ring closed}}} \right)$$

and

$$t_{\text{back-up relay}} = t_{\text{adjacent relay with ring closed}}$$

$$+ t_{\text{back-up relay with ring open}} \left(1 - \frac{t_{\text{remote relay with ring closed}}}{t_{\text{back-up relay with ring closed}}} \right)$$

$$t_{\text{back-up relay}} \geq t_{\text{relay next to fault}} + t_{\text{ discrimination margin}}$$

4. The same procedure is repeated for each relay, i.e. by considering a fault at its associated breaker terminals with the ring closed, and then for a fault at the opposite end of the line with the ring open. The procedure is completed when no further time dial setting changes are required.

It should be noted that calculating the time dial setting based on contact travel guarantees proper co-ordination of the relays on a ring, since the setting is carried out for the most severe conditions, i.e. for a fault at the various busbars with the ring closed, and with the ring open. To illustrate the above procedure, consider the system shown in Figure 7.16, with a 34.5 kV ring connecting three busbars.

After setting the pick-up and instantaneous current values, the steps to set the time dial values by considering the travel contact are as indicated below:

1. The time dial values of the relays are initially set in such a way that co-ordination is guaranteed with those relays associated with lines or machines supplied from the three busbars of the ring.
2. A fault at the breaker terminals associated with relay R_{21} with the ring closed is considered first. From Figure 7.17a, the three-phase short-circuit level is 5157 A.
3. The operating times of relays R_{13} and R_{14} should be checked to ensure that there is adequate co-ordination with relay R_{21}. If not, their time dial settings should be increased.
4. The operating times of relay R_{11} and its back-up R_{22} are calculated to ensure that the specified fault will be cleared with the ring closed.
5. The short-circuit current value for a fault at the breaker terminals associated with relay R_{11} is calculated, with the ring open. From Figure 7.17b, this value is 2471 A.
6. The operating times of relay R_{11} and its back-up relay R_{22} are calculated for this condition with the following expressions:

$$t''_{\text{relay } R_{11} \text{ next to fault}} = t_{R_{21} \text{ adjacent relay with ring closed}}$$

$$+ t'_{R_{11} \text{ relay next to fault with ring open}} \left(1 - \frac{t_{R_{21} \text{ adjacent relay with ring closed}}}{t_{R_{11} \text{ relay next to fault with ring closed}}} \right)$$

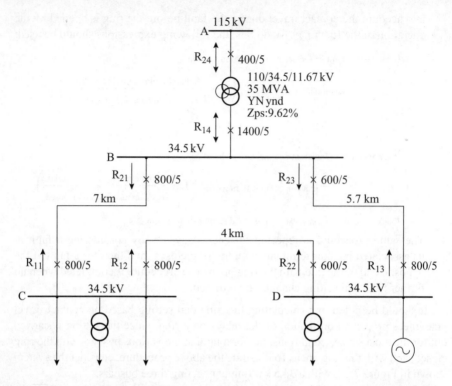

Figure 7.16   Ring system to illustrate overcurrent directional relay setting procedure

and

$$t''_{R_{22} \text{ back-up relay}} = t_{R_{21} \text{ remote relay with ring closed}}$$

$$+ t'_{R_{22} \text{ back-up relay with ring open}} \left(1 - \frac{t_{R_{21} \text{ remote relay with ring closed}}}{t_{R_{22} \text{ back-up relay with ring closed}}}\right)$$

where: t = operating time of the relay for the initial fault with the ring closed; t' = operating time of the relay with the new topology after the first relay operates; t'' = operating time of the relay considering the new topology and taking account of the contact travel.

7. Finally, the time dial setting of relay R_{22} should be checked to confirm that it satisfies the following expression:

$$t''_{R_{22}} \geq t''_{R_{11}} + t_{\text{discrim.margin}}$$

This procedure is then repeated for the rest of the relays of the ring. Table 7.1 summarises the steps as a guideline to completing this co-ordination exercise.

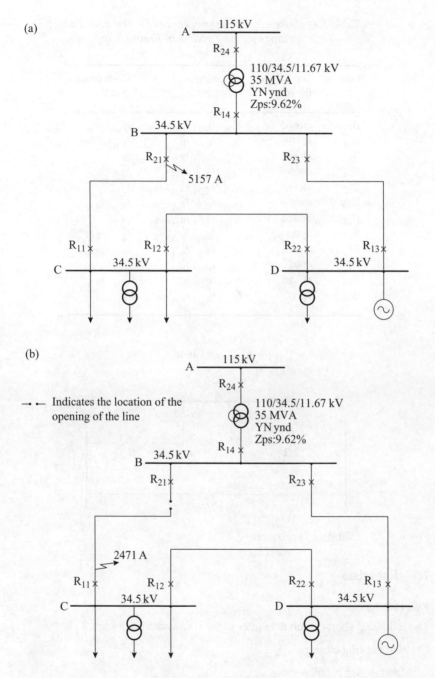

Figure 7.17 *Short-circuit currents for faults at breaker:* (a) fault associated with R_{21} with ring closed; (b) fault associated with R_{11} with ring open at R_{21}

Table 7.1   Summary of the procedures for the time dial settings for ring system in Figure 7.16

Fault at	Topology of the ring	Calculation	Co-ordination check
B_{21}	closed	$t_{R_{21}}$, $t_{R_{11}}$, $t_{R_{22}}$	$t_{R_{13}}$, $t_{R_{14}}$
B_{11}	open	$t'_{R_{11}}$, $t'_{R_{22}}$, $t''_{R_{11}}$	$t'_{R_{22}}$
B_{12}	closed	$t_{R_{12}}$, $t_{R_{22}}$, $t_{R_{23}}$	$t_{R_{21}}$
B_{22}	open	$t'_{R_{22}}$, $t'_{R_{23}}$, $t''_{R_{22}}$	$t''_{R_{23}}$
B_{13}	closed	$t_{R_{13}}$, $t_{R_{23}}$, $t_{R_{14}}$, $t_{R_{11}}$	$t_{R_{12}}$
B_{23}	open	$t'_{R_{23}}$, $t'_{R_{14}}$, $t'_{R_{11}}$, $t''_{R_{23}}$	$t''_{R_{14}}$, $t''_{R_{11}}$
B_{11}	closed	$t_{R_{11}}$, $t_{R_{21}}$, $t_{R_{14}}$, $t_{R_{13}}$	$t_{R_{22}}$
B_{21}	open	$t'_{R_{21}}$, $t'_{R_{14}}$, $t'_{R_{13}}$, $t''_{R_{21}}$	$t''_{R_{13}}$, $t''_{R_{14}}$
B_{22}	closed	$t_{R_{22}}$, $t_{R_{12}}$, $t_{R_{21}}$	$t_{R_{23}}$
B_{12}	open	$t'_{R_{12}}$, $t'_{R_{21}}$, $t'_{R_{12}}$	$t'_{R_{21}}$
B_{23}	closed	$t_{R_{23}}$, $t_{R_{13}}$, $t_{R_{12}}$	$t_{R_{14}}$, $t_{R_{11}}$
B_{13}	open	$t'_{R_{13}}$, $t'_{R_{12}}$, $t''_{R_{13}}$	$t''_{R_{12}}$

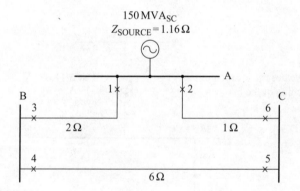

Figure 7.18   Network for Exercise 7.1

7.7   Exercises

Exercise 7.1

The following short-circuit data refer to the ring system in Figure 7.18:

With the ring closed:

fault at A = 6, 560 A
fault at B = 2, 731 A   2, 235 A (contribution from 1) plus 496 A from 2
fault at C = 2, 280 A   622 A (from 1) + 1, 658 A (from 2)

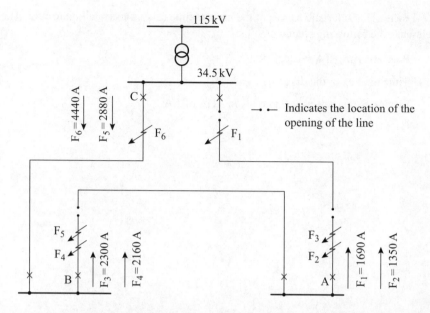

Figure 7.19   Network for Exercise 7.2. F_1, F_3 and F_5 correspond to faults with the ring open at the ends indicated by the symbol $\rightarrow \cdot \leftarrow$. F_2, F_4 and F_6 correspond to faults with the ring closed

With the ring open:

fault at B = 750 A   (with 1 open)
          = 2411 A   (with 2 open)
fault at C = 1832 A   (with 1 open)
          = 832 A   (with 2 open)

If the instantaneous element 6 at substation C is set to 424 A and protects up to the middle of the line A-C with the ring closed, calculate the setting for the instantaneous element of relay 4 at substation A.

Exercise 7.2

The short-circuit levels for different three-phase fault conditions on a 34.5 kV ring system are shown in Figure 7.19. The fault currents F_1, F_3 and F_5 take account of those lines that are open at the points indicated for each fault, and assume that each line is already open at the time the fault occurs.

Determine the relay time dial and instantaneous current settings for the overcurrent relays associated with breakers A, B and C. The tap of each relay was calculated beforehand and a value of 5 was selected for the three relays.

Use the characteristic curves of the inverse-time relay, shown in Figure 5.15. The relay has the following characteristics:

Pick-up: 1 to 12 A in steps of 1 A

Time dial: as in the diagram

Instantaneous settings: 6 to 80 A in steps of 1 A

Chapter 8
Differential protection

8.1 General

Differential protection functions when the vector difference of two or more similar electrical magnitudes exceeds a predetermined value. Almost any type of relay can function as differential protection – it is not so much the construction of the relay that is important but rather its method of connection in the circuit. The majority of the applications of differential relays are of the current-differential type, but they can also be of the voltage-differential type, operating on the same principle as the current relays; the difference lies in the fact that the operating signal is derived from a voltage across a shunt resistance.

A simple example of a differential arrangement is shown in Figure 8.1. The secondaries of the current transformers (CTs) are interconnected and the coil of an overcurrent relay is connected across these. Although the currents I_1 and I_2 may be different, provided that both sets of CTs have appropriate ratios and connections then, under normal load conditions or when there is a fault outside the protection

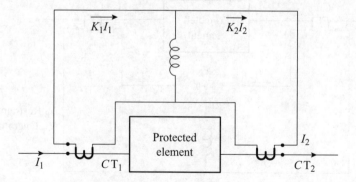

Figure 8.1   Differential protection – current balance

zone of the element, the secondary currents will circulate between the two CTs and will not flow through the overcurrent relay. However, if a fault occurs in the section between the two CTs the fault current would flow towards the short-circuit from both sides and the sum of the secondary currents would flow through the differential relay. In all cases the current in the differential relay would be proportional to the vector difference between the currents that enter and leave the protected element; if the current through the differential relay exceeds the setting value then the relay will operate.

One arrangement that is extensively used is the differential relay with a variable-percentage characteristic, alternatively described as being percentage biased, which has an additional unit, the restraining coil, in addition to the operating coil as shown in Figure 8.2. Since this type of relay is most common in differential relaying schemes, all references to differential relays in this chapter will be on the basis of using this type of relay. The current in the operating coil is proportional to $(I_1 - I_2)$. If N is equal to the number of turns of the restraining coil with the operating coil connected to the mid-point of the restraining coil, then the total ampere-turns are equal to $I_1(N/2)$ plus $I_2(N/2)$, which is the same as if $(I_1 + I_2)/2$ flowed through all of the restraining coil. The operating characteristic of the relay with this form of restraint is shown in Figure 8.3.

CTs, although produced to the same specification, will not have identical secondary currents for the same primary currents because of slight differences in their magnetising characteristics. The relay restraining force increases with the magnitude of $(I_1 + I_2)$, thus preventing unnecessary tripping due to any CT unbalance errors. In addition, the restraining torque is increased in the presence of through-fault currents producing a more stable operating characteristic and preventing relay mal-operation. In relays that have variable tappings in the restraint coil circuits, the tappings can be set to balance out any currents due to differences in the CTs. If the relays do not have these variable tappings, then the currents leaving the CTs should match closely in order to avoid mal-operation of the relays.

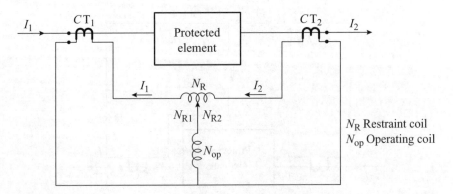

Figure 8.2   Differential relay with variable-percentage characteristic

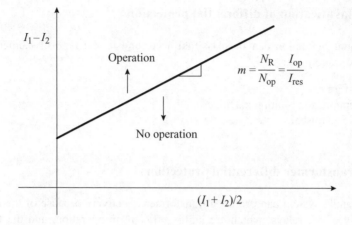

Figure 8.3   Relay characteristic (variable-percentage type)

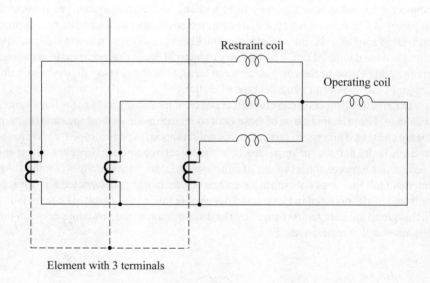

Element with 3 terminals

Figure 8.4   Differential protection for element with three terminals

Differential relays can be used for power system elements that have more than two terminals, as shown in Figure 8.4. Each of the three restraining coils has the same number of turns and each coil produces a restraint that is independent of the others; these add arithmetically. The slope of the operating characteristic of each relay varies depending on the current distribution in the three restraining coils. There are also other types of differential relays which use directional or overvoltage elements in place of the overcurrent elements. Thus, all types are extensions of the fundamental principles that have been described above.

8.2   Classification of differential protection

Differential protection can be classified according to the type of element to be protected, as follows:

- transformers;
- generators and rotating machines;
- lines and busbars.

8.3   Transformer differential protection

A differential system can protect a transformer effectively because of the inherent reliability of the relays, which are highly efficient in operation, and the fact that equivalent ampere-turns are developed in the primary and secondary windings of the transformer. The CTs on the primary and secondary sides of the transformer are connected in such a way that they form a circulating current system, as illustrated in Figure 8.5. Faults on the terminals or in the windings are within the transformer protection zone and should be cleared as quickly as possible in order to avoid internal stress and the danger of fire. The majority of internal faults that occur in the windings are to earth (across to the core) or between turns, with the severity depending on the design of the transformer and the type of earthing.

Differential protection can also detect and clear insulation faults in the transformer windings. The principal cause of these faults is arcing inside the bushings and faults in the tap changer. This type of protection not only responds to phase-to-phase and phase-to-earth faults but also in some degree to faults between turns. However, phase-to-phase faults between the windings of a three-phase transformer are less common. An internal fault that does not constitute an immediate danger is designated an incipient fault and, if not detected in time, could degenerate into a major fault. The main faults in this group are core faults, caused by the deterioration of the insulation between the laminations that make up the core.

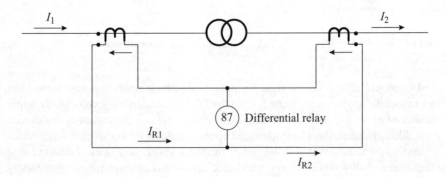

Figure 8.5   Transformer differential protection

8.3.1   Basic considerations

In order to apply the principles of differential protection to three-phase transformers, the following factors should be taken into account:

Transformation ratio

The nominal currents in the primary and secondary sides of the transformer vary in inverse ratio to the corresponding voltages. This should be compensated for by using different transformation ratios for the CTs on the primary and secondary sides of the transformer.

Transformer connections

When a transformer is connected in star/delta, the secondary current has a phase shift of a multiple of 30° relative to the primary, depending on the vector group. This shift can be offset by suitable secondary CT connections. Furthermore, the zero-sequence current that flows in the star side of the transformer will not induce current in the delta winding on the other side. The zero-sequence current can therefore be eliminated from the star side by connecting the CTs in delta. For the same reason, the CTs on the delta side of the transformer should be connected in star. When CTs are connected in delta, their nominal secondary values should be multiplied by $\sqrt{3}$ so that the currents flowing in the delta are balanced by the secondary currents of the CTs connected in star.

Tap changer

If the transformer has the benefit of a tap changer it is possible to vary its transformation ratio, and any differential protection system should be able to cope with this variation. Since it is not practical to vary the CT transformation ratios, the differential protection should have a suitable tolerance range in order to be able to modify the sensitivity of its response of operation. For this reason it is necessary to include some form of biasing in the protection system together with some identifying markings of the higher current input terminals.

Magnetisation inrush

This phenomenon occurs when a transformer is energised, or when the primary voltage returns to its normal value after the clearance of an external fault. The magnetising inrush produces a current flow into the primary winding that does not have any equivalent in the secondary winding. The net effect is thus similar to the situation when there is an internal fault on the transformer. Since the differential relay sees the magnetising current as an internal fault, it is necessary to have some method of distinguishing between the magnetising current and the fault current. These methods include:

1. Using a differential relay with a suitable sensitivity to cope with the magnetising current, usually obtained by a unit that introduces a time delay to cover the period of the initial inrush peak.

2.   Using a harmonic-restraint unit, or a supervisory unit, in conjunction with a differential unit.
3.   Inhibiting the differential relay during the energising of the transformer.

8.3.2   Selection and connection of CTs

The following factors should be taken into account when considering the application of differential protection systems:

1.   In general, the CTs on the star side of a star/delta transformer should be connected in delta, and those on the delta side should be connected in star. This arrangement compensates for the phase shift across the transformer and blocks the zero-sequence current in the event of external faults to earth.
2.   The relays should be connected to accept the load current entering one side of the transformer and leaving by the other side. If there are more than two windings it is necessary to consider all combinations, taking two windings at a time.
3.   The CT ratios should be selected in order to produce the maximum possible balance between the secondary currents of both sides of the transformer under maximum load conditions. If there are more than two windings, all combinations should be considered, taking two windings at a time and the nominal power of the primary winding. If the available CT ratios do not enable adequate compensation to be made for any variation in secondary current from CTs, then compensation transformers can be used to offset the phase shift across the transformer.

The following examples show the connections of the CTs, the calculation of their transformation ratios, and the connection of the differential relays as applied to transformer protection schemes.

Example 8.1

Consider a 30 MVA, 11.5/69 kV, Yd1 transformer as shown in the single-line diagram in Figure 8.6. Determine the transformation ratio and connections of the CTs required in order to set the differential relays. CTs with ratios in steps of 50/5 up to 250/5, and in steps of 100/5 thereafter, should be used. Use relays with a variable-percentage

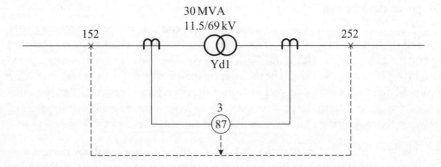

Figure 8.6   Single-line diagram for Example 8.1

characteristic. The available current taps are: 5.0-5.0, 5.0-5.5, 5.0-6.0, 5.0-6.6, 5.0-7.3, 5.0-8.0, 5.0-9.0, and 5.0-10.0 A.

Solution

Figure 8.7 shows the complete schematic of the three-phase connections. The currents in the windings and in the lines are drawn and show that the restraint currents on the star and delta sides of the relay are in phase.

For a throughput of 30 MVA the load currents are

$$I_{load}(69\,kV) = \frac{30\,MVA}{\sqrt{3} \times 69\,kV} = 251.0\,A$$

$$I_{load}(11.5\,kV) = \frac{30\,MVA}{\sqrt{3} \times 11.5\,kV} = 1506.13\,A$$

In order to increase the sensitivity, the CT ratio at 11.5 kV is selected as close as possible to the maximum load current and, therefore, a CT ratio (11.5 kV) of

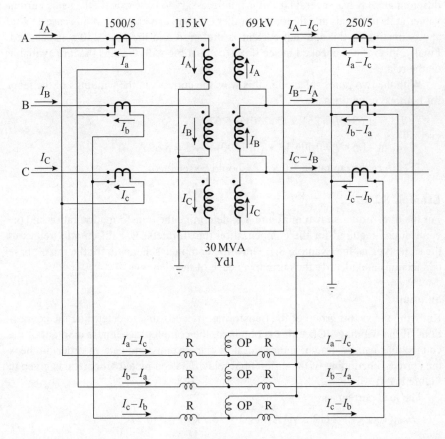

Figure 8.7   Three-phase connection diagram for Example 8.1

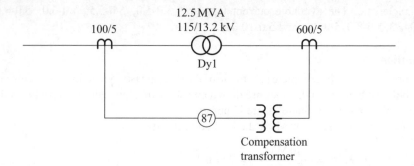

Figure 8.8   Single-line diagram for Example 8.2

1500/5 is chosen. When calculating the other CT ratio, a balance of currents has to be achieved, i.e. $1506.13 \times (5/1500) \times \sqrt{3} = 251 \times (5/X)A \Rightarrow X = 144$. This would suggest using a CT ratio of 150/5. However, taking into account the fact that the differential relay has several taps, it is not necessary to have exactly the same current values at its terminals and therefore another CT ratio can be used. In this case a value approximating to the nominal current is preferred and the ratio 250/5 is selected. Finally, this ratio is checked to see if it is compatible with the taps that are available on the relay.

With the two ratios selected in this way, the currents in the windings of the relay for nominal conditions are

$$I_{relay} \text{ at } 69\,kV = 251 \times (5/250) = 5.02\,A$$

$$I_{relay} \text{ at } 11.5\,kV = 1506.13 \times (5/1500) \times \sqrt{3} = 8.69\,A$$

Therefore the tap range 5.0-9.0 A should be selected.

Example 8.2

For the transformer shown in Figure 8.8, determine the transformation ratios and the connections required for the compensation transformers. Use differential relays of the same type as for Example 8.1. Draw the complete schematic for the three-phase connections and identify the currents in each of the elements.

Solution

Knowing the vector group of the transformer it is possible to determine the connections of the windings. Once this is obtained, the complete schematic diagram of the connections can be drawn, indicating how the currents circulate in order to check the correct functioning of the differential relay. The completed schematic is given in Figure 8.9.

The load currents are

$$I_{load}(13.2\,kV) = (12.5\,MVA)/(\sqrt{3} \times 13.2\,kV) = 546.7\,A$$

$$I_{load}(115\,kV) = (12.5\,MVA)/(\sqrt{3} \times 115\,kV) = 62.75\,A$$

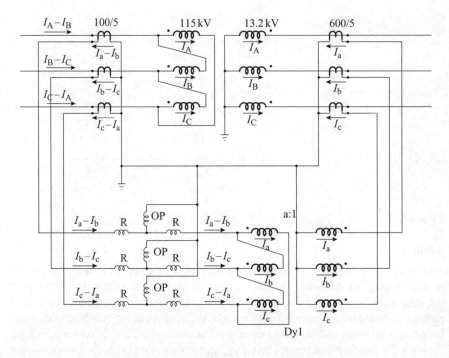

Figure 8.9   Three-phase connection diagram for Example 8.2

In order to select the ratio of the compensation transformers it is necessary to obtain a correct balance of currents, taking into account the delta connection on one of the sides, i.e. $62.75\,\mathrm{A} \times (5/100) = 546.7\,\mathrm{A} \times (5/600) \times (1/a) \times \sqrt{3}$, from which $a = 2.51$. However, let us assume that this value of a is not included on commercially available relays but that a value of $a = \sqrt{3}$ is, this being a very typical value provided on this type of relay. On this basis the currents feeding into the relay are:

$$I_{\mathrm{relay}}(115\,\mathrm{kV}) = 62.75 \times (5/100) = 3.13\,\mathrm{A}$$

$$I_{\mathrm{relay}}(13.2\,\mathrm{kV}) = 546.7 \times (5/600) \times (1/\sqrt{3}) \times \sqrt{3} = 4.55\,\mathrm{A}$$

Selection of the tap setting:

$$\frac{4.55}{3.13} = \frac{X}{5},$$

from which $X = 7.26\,\mathrm{A}$, so that for these conditions the tap to be selected is 5.0-7.3 A.

Example 8.3

In the system in Figure 8.10, given the power rating and voltage ratio of the power transformer and the CT ratios, determine the transformation ratios and the connections of the compensation transformers required in order to set the differential relays, which are not provided with taps. Produce the complete three-phase connection schematic and identify the currents in each of the elements.

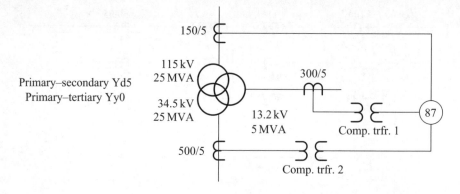

Figure 8.10   Single-line diagram for Example 8.3

Solution

The three-phase diagram of connections is given in Figure 8.11; the connections of the windings of the power transformer can be obtained from the vector group. If the differential relays are connected on the Y side, then compensation transformer number 2 should be connected in Yd5 in order to compensate for the phase difference between the primary and secondary currents. There is no need for phase compensation between the primary and tertiary winding, and a Yy0 connection for compensation transformer 1 is therefore the most appropriate arrangement. It should be noted that the common point of the operating coils and the neutral points of the compensation transformers and CTs must be connected to only one earthing point in order to avoid mal-operation during external faults. If several earthing points are used circulating currents could be induced during external faults causing the relays to pick up inadvertently.

When determining the ratios of the compensation transformers, the calculations involving the primary and secondary windings should be carried out on the basis of the main transformer having only two windings with no current circulating in the tertiary. The calculations involving the primary and tertiary windings should be treated in a similar way. This method ensures that a correct selection is made which will cover any fault or load current distribution.

Considering the currents on the primary and secondary sides, $I_{relay}(115\,kV) = [25\,MVA/(\sqrt{3} \times 115\,kV)] \times (5/150) = 125.51 \times (5/150) = 4.18\,A$. The current in the relay associated with the 34.5 kV side should be equal to 4.18 A, so that

$$I_{relay}(34.5\,kV) = 4.18\,A = \frac{25\,MVA}{\sqrt{3} \times 34.5\,kV} \times \frac{5}{500} \times \frac{1}{\sqrt{3}} \times \frac{1}{a_2}$$

from which $a_2 = 0.578$. The turns ratio of compensator 2 is therefore $1/0.578 = 1.73$, provided that the delta side has the higher number of turns.

The current in the relay for the 13.2 kV side is calculated assuming that the power of the tertiary winding is equal to that of the primary winding, thus obtaining a correct balance of magnitudes. This connection is equivalent to taking the primary and the tertiary and treating them as a transformer with two windings so

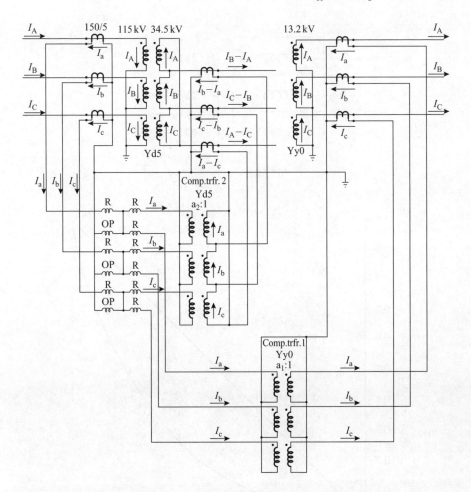

Figure 8.11   Three-phase connection diagram for Example 8.3

that $I_{relay}(13.2\,\mathrm{kV}) = 4.18\,\mathrm{A} = (25\,\mathrm{MVA}/(\sqrt{3} \times 13.2\,\mathrm{kV})) \times (5/300) \times (1/a_1)$, from which $a_1 = 4.36$. With this setting, it can be shown that, for any load distribution, the primary restraint current is equal to the sum of the secondary and tertiary restraint currents.

8.3.3   Percentage of winding protected by the differential relay during an earth fault

Although differential protection is very reliable for protecting power transformers, the windings are not always fully protected, especially in the case of single-phase faults. Consider the case of a delta/star transformer as shown in Figure 8.12a, in which the star winding has been earthed via a resistor. Assume an internal earth fault occurs at point F at a distance X from the neutral point, involving X per cent turns,

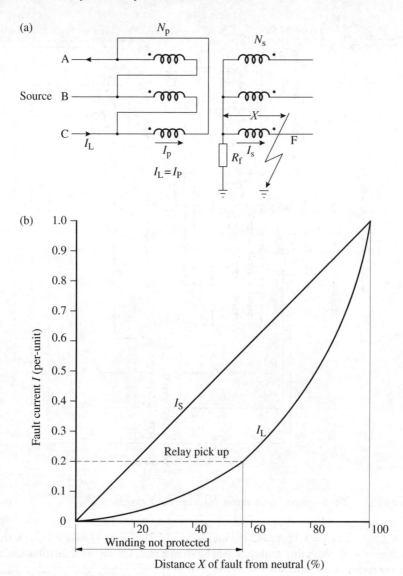

Figure 8.12   Delta/star transformer; star winding earthed via a resistor, with a fault on the star side: (a) connection diagram; (b) fault current values for primary and secondary

and that the resistor has been set so that nominal current I_{nom} will flow for a fault on the terminals, (with full line-to-neutral voltage applied between phase and earth). The numbers of primary and secondary turns are N_p and N_s respectively.

The secondary current for a fault at F is produced by X per cent of the line-to-neutral voltage. Therefore, by direct ratio, the current will be $X I_{nom}$. In addition, the number of turns involved in the fault is XN_s. The distribution of current in the delta side for an earth fault on the star side results in a line current I'_L equal to the phase current. Therefore

$$I'_L = XI_{nom} \times (XN_s/N_p) = X^2 I_{nom}(N_s/N_p) \tag{8.1}$$

Under normal conditions, the line current in the delta side, I_L, is

$$I_L = \sqrt{3}I_{nom} \times (N_s/N_p) \tag{8.2}$$

If the differential relay is set to operate for 20 per cent of the nominal line current then, for operation of the relay, the following should apply:

$$I'_L \geq 0.2 \times I_L$$

i.e.

$$X^2 I_{nom}(N_s/N_p) \geq 0.2 \times \sqrt{3} \times I_{nom} \times (N_s/N_p)$$

$$X^2 \geq 0.2\sqrt{3}, \quad \text{i.e. } X \geq 59 \text{ per cent}$$

Therefore, 59 per cent of the secondary winding will remain unprotected. It should be noted that to protect 80 per cent of the winding ($X \geq 0.2$) would require an effective relay setting of 2.3 per cent of the nominal primary current. This level of setting can be very difficult to achieve with certain types of differential relays.

Figure 8.12b illustrates typical primary and secondary currents for delta/star transformers, where the secondary star winding is earthed via a resistor, and also the effect of the location of a fault along the star winding on the pick up of the differential relays.

8.3.4 Determination of the slope

The setting of the slope of differential relays is carried out with the aim of ensuring that there will be no mal-operations because of differences in the currents in the restraint windings due to the transformation ratios of the CTs and the operation of the tap changer under load conditions. In order to determine the slope, the restraint and operating torques are calculated on the basis of the currents and the number of turns in the respective coils as set out below:

$$T_{res} = I_1 N_{R1} + I_2 N_{R2}$$

$$T_{op} = |I_1 - I_2| N_{op}$$

where: $I_1, I_2 =$ currents in the secondaries of the CTs; $N_{R1}, N_{R2} =$ number of turns on the restraint coils (see Figure 8.2).

In order for the relay to operate, $T_{op} > T_{res}$, i.e. $|I_1 - I_2| N_{op} > I_1 N_{R1} + I_2 N_{R2}$. If $N_{R1} = N_{R2} = N_R/2$, this then gives $T_{res} = (I_1 + I_2)N_R/2$.

For relay operation, the slope

$$\frac{|I_1 - I_2|}{0.5|I_1 + I_2|} \geq \frac{N_{res}}{N_{op}} = m$$

A typical operating curve is shown in Figure 8.3.

8.3.5   Distribution of fault current in power transformers

When considering the operation of differential protection it is very important to take into account the distribution of fault current in all the windings in order to ensure that the settings that have been selected have a suitable sensitivity. This is particularly critical for single-phase faults in transformers that are earthed via an impedance. The following example illustrates the procedure.

Example 8.4

A 115/13.2 kV Dy1 transformer rated at 25 MVA has differential protection as indicated in Figure 8.13. The transformer is connected to a radial system, with the source on the 115 kV side. The minimum operating current of the relays is 1 A. The transformer 13.2 kV winding is earthed via a resistor which is set so that the current for a single-phase fault on its secondary terminals is equal to the nominal load current.

Draw the complete three-phase diagram and indicate on it the current values in all the elements for:

(i)   Full load conditions.
(ii)   When a fault occurs at the middle of the winding on phase C, on the 13.2 kV side, assuming that the transformer is not loaded.

For both cases indicate if there is any relay operation.

Solution

Full load conditions

The full load conditions for the maximum load of the transformer are as follows:

$$I_{nom(13.2\,kV)} = \frac{25 \times 10^6 \text{ VA}}{\sqrt{3} \times 13.2 \times 10^3 \text{ V}} = 1093.47 \text{ A}$$

and

$$I_{nom(115\,kV)} = \frac{25 \times 10^6 \text{ VA}}{\sqrt{3} \times 115 \times 10^3 \text{ V}} = 125.51 \text{ A}$$

It should be noted that, based on the primary currents given, the phase rotation A-B-C is negative, i.e. clockwise. Therefore, the respective currents at the secondary lead the primary currents by 30° in order to provide the required phase shifting for the transformer vector group (Dy1). Figure 8.13 shows the current values through the HV and LV connections, and it is clear that balanced currents are presented to the differential relays, which therefore do not pick up, as expected.

Fault at the middle of 13.2 kV winding C

Since the transformer is earthed through a resistor that limits the current for faults at the transformer 13.2 kV bushings to the rating of the winding, and since the fault is at the middle of the winding, the fault current is then equal to half the rated value as follows:

$$I_{fault} = (I_{nom(13.2\,kV)})/2 = 1093.47/2 = 546.74 \text{ A}$$

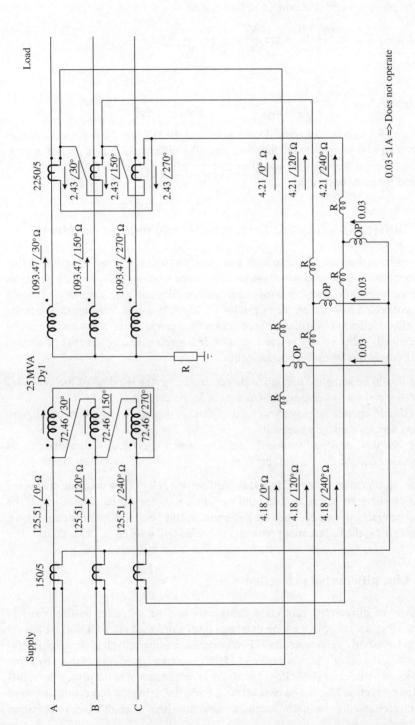

Figure 8.13 Three-phase connection diagram for Example 8.4

The primary current within the delta winding is

$$I_{prim} = I_{fault} \times \frac{(N_2/2)}{N_1}, \text{since } \frac{N_2}{N_1} = \frac{V_2}{\sqrt{3} \times V_1}$$

Then

$$I_{prim} = I_{fault} \times \frac{V_2}{2 \times \sqrt{3} \times V_1} = 546.47 \times \frac{13.2}{2 \times \sqrt{3} \times 115} = 18.12\,\text{A}$$

Figure 8.14 shows the current values through the HV and LV connections, from which it can be seen that, for this case also, the differential relays do not operate since the current through their operating coils is only 0.6 A, which is less than the 1 A required for relay operation.

8.4 Differential protection for generators and rotating machines

Differential protection for generators and other rotating machines is similar to that for transformers in several ways. Internal generator winding faults include phase-to-phase short-circuits, short-circuited turns, open circuits and faults to earth, and should be disconnected by opening the circuit as quickly as possible. In order to obtain the most effective form of differential protection the neutral of the generator should be well earthed, either solidly or via a resistor or a reactor. The differential protection should satisfy the following requirements:

1. It should be sensitive enough to detect damage in the winding of the generator stator, and yet not operate for faults outside the machine.
2. It should operate quickly in such a way that the generator is disconnected before any serious damage can result.
3. It should be designed so that the main breaker is opened, as well as the neutral breaker and the field circuit breaker.

The arrangements of the CTs and the differential relays for a machine connected in star can be seen in Figure 8.15, and in Figure 8.16 for a delta connection. If the neutral connection is made inside the generator and the neutral taken outside as shown in Figure 8.17, the differential protection provided will only cover earth faults.

8.5 Line differential protection

The form of differential protection using only one set of relays as illustrated in Figure 8.2 is not suitable for long overhead lines since the ends of a line are too far apart to be able to interconnect the CT secondaries satisfactorily. It is therefore necessary to install a set of relays at each end of the circuit and interconnect them by some suitable communication link. Pilot protection is an adaptation of the principles of differential protection that can be used on such lines, the term pilot indicating that there is an interconnecting channel between the ends of the lines through which information

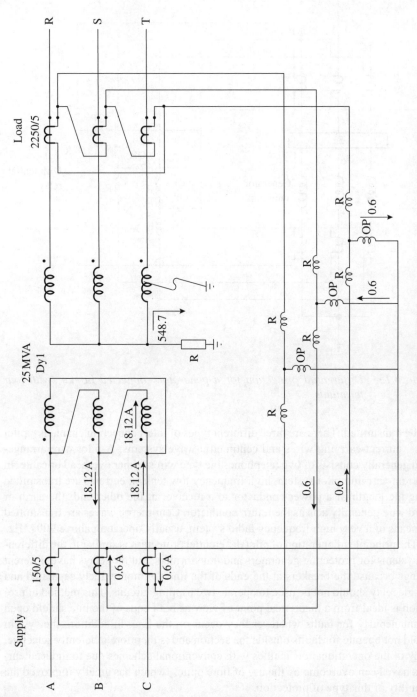

Figure 8.14 Conditions for a fault at the middle of the winding on phase C on the 13.2 kV side (Example 8.4)

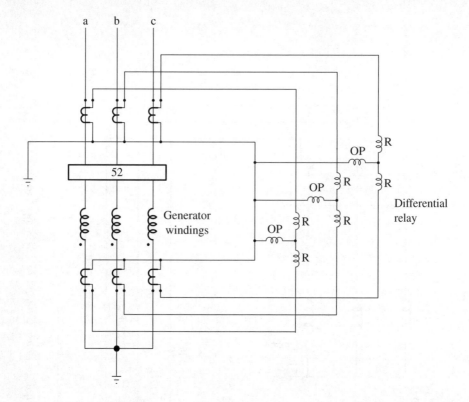

Figure 8.15   Differential protection for a generator connected in star, with four terminals

can be transmitted. There are three different types of interconnecting channels – pilot wires, current-carrying wires and centimetric-wave systems. A pilot wire arrangement generally consists of two telephone-line type wires, either overhead or cable. In a current-carrying pilot system, high frequency low tension currents are transmitted along the length of a power conductor to a receiver at the other end; the earth or guard wire generally acts as the return conductor. Centimetric waves are transmitted by means of a very high frequency radio system, usually operating above 900 MHz.

The principle of operation of pilot differential protection is similar to the differential systems for protecting generators and transformers, but the relays have different settings because the breakers at the ends of the line are more widely separated and a single relay should not be used to operate two tripping circuits. This method of protection is ideal from a theoretical point of view as both ends of the line should open instantaneously for faults wherever they occur on the line. In addition, the system should not operate for faults outside the section and is therefore inherently selective. Many of the operational difficulties with conventional schemes due to induced currents have been overcome by the use of fibre optics, which has greatly improved the reliability of this type of protection.

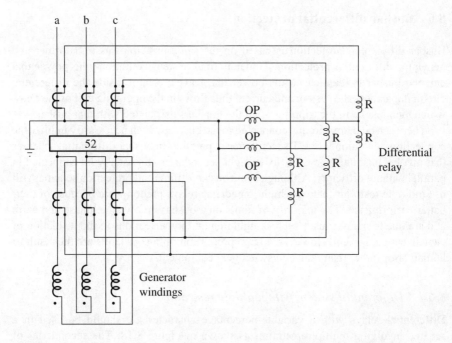

Figure 8.16   Differential protection for a generator connected in delta

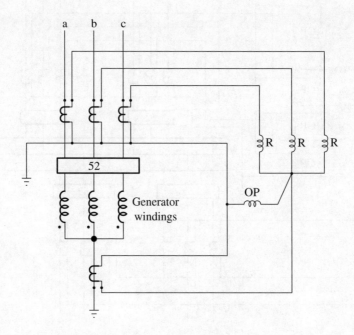

Figure 8.17   Differential protection for a generator connected in star

8.6   Busbar differential protection

Busbar differential protection is based on the same principles as transformer and generator differential protection. Under normal system conditions the power that enters a busbar is identical to the power that leaves; a fault inside the differential circuit unbalances the system and current thus flows in the operating coil of the relay, which then results in the tripping of all the breakers associated with that busbar.

There can be many circuits connected to the busbar, which necessarily implies the connection of a number of CT secondaries in parallel. In busbar differential schemes that involve bushing type CTs, six to eight secondaries can usually be connected in parallel without difficulty. Although some busbar differential protection schemes still use multiple restraint features, high impedance relays predominate because of their better performance. The majority of faults on a busbar involve one phase and earth and are due to many causes such as lightning and imperfections in the insulation of switchgear equipment. However, a large proportion of busbar faults are the result of human error rather than faults on switchgear equipment.

8.6.1   *Differential system with multiple restraint*

Differential relays with a variable-percentage characteristic should be used in a scheme involving multiple restraint, as shown in Figure 8.18. The secondaries of

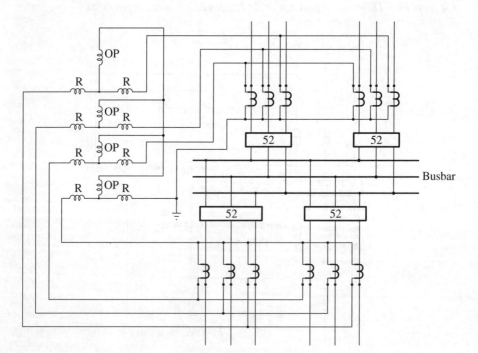

Figure 8.18   Multiple restraint busbar differential protection

the CTs on the feeders on the outgoing side of the busbar are connected in parallel and across the differential relay, together with the secondaries of CTs of the circuits on the incoming side of the busbar which also are connected in parallel.

8.6.2 High impedance differential system

The high impedance arrangement tends to force any incorrect differential current to flow through the CTs instead of through the operating coil of the relay, and thus avoids mal-operation for external fault or overload conditions when the secondary currents of all the CTs are not the same because of differences in the magnetisation characteristics.

Connecting CTs in parallel

This arrangement requires only one high impedance relay, connected across the terminals of the CT secondaries which are connected in parallel to a set of CTs per circuit, as shown in Figure 8.19. However, with the connections made in this way, the busbar is only protected from earth faults. In order for the scheme to be effective, the resistance of the CT secondary wiring should be as low as possible.

The relay basically consists of an instantaneous overvoltage unit which is set by calculating the maximum voltage at the relay terminals for an external fault, taking into account the maximum primary fault current, the resistance of the secondary windings and the wiring, and the transformation ratios of the CTs, plus a safety margin. Consequently, during an external fault, the voltage across the terminals of the relay is relatively low and does not initiate any relay operation. During internal faults the voltage across the relay terminals is higher and results in the operation of the instantaneous overvoltage unit which sends a tripping signal to the appropriate breakers.

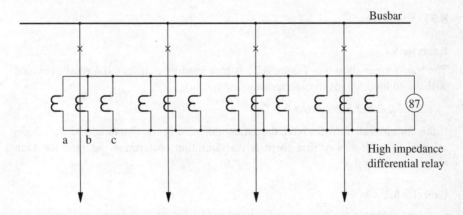

Figure 8.19   High impedance differential protection scheme with CTs in parallel

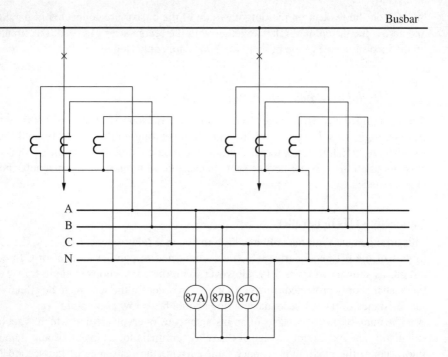

Figure 8.20   High impedance differential protection arrangement using a common bus for each phase

Using common buses for each phase

In this scheme balanced groups of CTs are formed in each one of the phases, using a common bus to feed three elements of a high impedance differential relay as illustrated in Figure 8.20, so that the busbar is protected against both phase and earth faults.

8.7   Exercises

Exercise 8.1

The transformer shown in Figure 8.21 is protected by a differential relay provided with the following taps in each restraint circuit:

2.9-3.2-3.5-3.8-4.2-4.6-5.0-8.7

If the current transformers have the ratios indicated in the diagram, and have been selected in such a way that there is no saturation, determine the taps for each restraint coil.

Exercise 8.2

Carry out the same calculations as for Exercise 8.1 for the transformer in Figure 8.22, using a differential relay with the same taps for each restraint circuit.

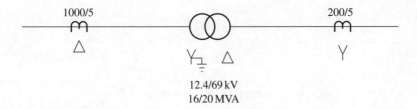

Figure 8.21 Diagram for Exercise 8.1

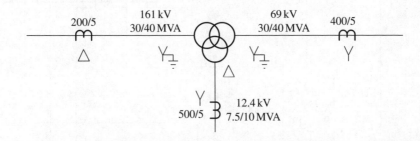

Figure 8.22 Diagram for Exercise 8.2

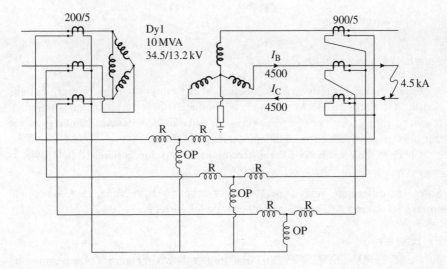

Figure 8.23 Diagram for Exercise 8.3

Exercise 8.3

Consider the 10 MVA, 34.5/13.2 kV, Dy1 transformer whose connections are given in Figure 8.23. The CTs on the 34.5 kV side of the transformer have a ratio of 200/5, and those on the 13.2 kV side a ratio of 900/5.

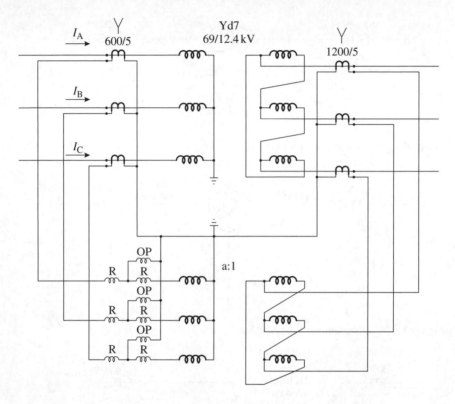

Figure 8.24   Diagram for Exercise 8.4

Calculate the magnitude and direction of the currents in the primary circuits and the CT secondaries. In addition, determine if these result in the differential relays operating for a fault between phases *b* and *c* in the secondary windings of the transformer which are carrying a fault current of 4.5 kA.

Check if the operation of the differential relays is appropriate. If the answer is negative, indicate what correction should be applied.

Note: The differential relays operate for a current equal to 20 per cent of nominal current, which is 5 A. Therefore, they operate for currents above 1 A.

Exercise 8.4

For the 16/20 MVA, 69/12.4 kV, Yd7 transformer shown in Figure 8.24, determine the transformation ratios and connections of the compensation transformers in order to connect the differential relays, which are not provided with taps to vary the settings. The primary and secondary windings of the compensators can only be connected in star or delta.

Draw the three-phase schematic diagram of the connections and identify the currents in each of the elements of the system, taking into account that I_A, I_B and I_C feed into the transformer primary.

Chapter 9

Distance protection

9.1 General

It is essential that any faults on a power system circuit are cleared quickly; otherwise they could result in the disconnection of customers, loss of stability in the system and damage to equipment. Distance protection meets the requirements of reliability and speed needed to protect these circuits, and for these reasons is extensively used on power system networks.

Distance protection is a nonunit type of protection and has the ability to discriminate between faults occurring in different parts of the system, depending on the impedance measured. Essentially, this involves comparing the fault current, as seen by the relay, against the voltage at the relay location to determine the impedance down the line to the fault. For the system shown in Figure 9.1, a relay located at A uses the

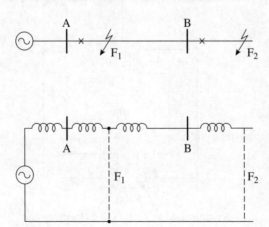

Figure 9.1   Faults occurring on different parts of a power system

line current and the line voltage to evaluate $Z = V/I$. The value of the impedance Z for a fault at F_1 would be Z_{AF1}, and $(Z_{AB} + Z_{BF2})$ for a fault at F_2.

The main advantage of using a distance relay is that its zone of protection depends on the impedance of the protected line that is a constant virtually independent of the magnitudes of the voltage and current. Thus, the distance relay has a fixed reach, in contrast to overcurrent units where the reach varies depending on system conditions.

9.2   Types of distance relays

Distance relays are classified depending on their characteristics in the R-X plane, the number of incoming signals and the methods used to compare the incoming signals. The most common type compares the magnitude or phase of the two incoming signals in order to obtain the operating characteristics, which are straight or circular lines when drawn in the R-X plane. Any type of characteristic obtainable with one type of comparator can also be obtained with the other, although the quantities compared would be different in each case.

If Z_R is the impedance setting of the distance relay, it should operate when $Z_R \geq V/I$, or when $I Z_R \geq V$. As shown in Figure 9.2, this condition can be obtained in the amplitude comparator that operates when the ampere-turns of the current circuit are greater than the ampere-turns of the voltage circuit. However, it is difficult to provide an amplitude comparator that functions correctly under fault conditions when the phase displacement between V and I tends to be 90° and transients are present, which leads to incorrect r.m.s. values of V and I which are required to evaluate $I Z_R \geq V$. For these reasons, the use of amplitude comparators is limited; it is more convenient to compare two signals by their phase difference than by their amplitudes.

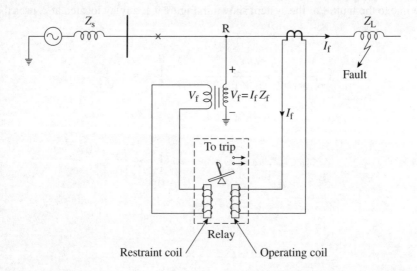

Figure 9.2   Relay based on an amplitude comparator

The following analysis shows that for two signals, S_0 and S_r, which are to be compared in magnitude, there exist two other signals S_1 and S_2 that can be compared by phase. The relationship between the signals is as follows:

$$S_0 = S_1 + S_2$$
$$S_r = S_1 - S_2$$

(9.1)

From eqns. 9.1

$$S_1 = \frac{(S_0 + S_r)}{2}$$
$$S_2 = \frac{(S_0 - S_r)}{2}$$

(9.2)

The comparison of the amplitudes is given by:

$$|S_0| \geq |S_r|$$
$$|S_1 + S_2| \geq |S_1 - S_2|$$

(9.3)

Defining $S_1/S_2 = C$, the relationship 9.3 can be expressed as

$$|C + 1| \geq |C - 1|$$

(9.4)

Drawing C in the R-X plane, as shown in Figure 9.3, it can be seen that condition 9.4 is satisfied in the semiplane on the right. This semiplane is defined for all the points $C\angle\Theta$ in such a way that $-90° \leq \Theta \leq +90°$.

Given that $C\angle\Theta = (S_1\angle\alpha)/(S_2\angle\beta)$, then the relationship 9.4 is satisfied when

$$-90° \leq \alpha - \beta \leq +90°$$

(9.5)

The above relationships demonstrate that two signals, obtained for use with an amplitude comparator, can be converted in order to be used by a phase angle comparator. The signals to be compared are analysed in the following paragraphs to obtain the operating characteristics of the main types of distance relays.

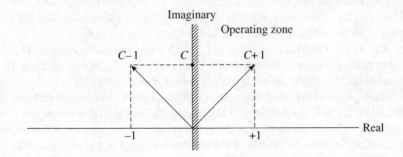

Figure 9.3   Comparison of phases in a complex plane: $C = S_1/S_2$

9.2.1   *Impedance relay*

The impedance relay does not take into account the phase angle between the voltage and the current applied to the relay and, for this reason, its operating characteristic in the R-X plane is a circle with its centre at the origin of the co-ordinates and a radius equal to the setting in ohms. The relay operates for all values of impedance less than the setting, i.e. for all the points inside the circle. Thus, if Z_R is the impedance setting, it is required that the relay will operate when $Z_R \geq V/I$, or when $I Z_R \geq V$. In order for an impedance relay to work as a phase comparator, the following signals should be assigned to S_0 and S_r:

$$S_0 = I Z_R$$
$$S_r = K V \tag{9.6}$$

The constant K takes into account the transformation ratios of the CTs and VTs. The corresponding signals for a phase comparator are

$$S_1 = K V + I Z_R$$
$$S_2 = -K V + I Z_R \tag{9.7}$$

Dividing eqns. 9.7 by KI, gives

$$S_1 = Z + Z_R/K$$
$$S_2 = -Z + Z_R/K \tag{9.8}$$

where

$$Z = V/I.$$

Note that the magnitudes of signals S_1 and S_2 have been changed when dividing by KI. However, this is not important since the main purpose is to retain the phase difference between them. It should be noted that drawing S_1 and S_2 in one or the other scale does not affect the phase relationship between the two signals.

Drawing Z_R/K and the eqns. 9.8 in the R-X plane, the operating characteristic of the relay is determined by the locus of the points Z such that Θ, the phase angle between S_1 and S_2, is given by $-90° \leq \Theta \leq +90°$. The construction is shown in Figure 9.4. Eqns. 9.8 give the origin of the rhomboid OABC which has diagonals of S_1 and S_2. From the properties of the rhomboid, the angle between S_1 and S_2 is 90° if $|Z| = |Z_R/K|$. Therefore point C is the limit of the operating zone, and the locus of point C for the different values of Z is a circle of radius Z_R/K.

If $Z < Z_R/K$, the situation shown in Figure 9.5 is obtained. In this case Θ is less than 90° and consequently the vector for Z is inside the relay operating zone. If, on the other hand, $Z > Z_R/K$, as in Figure 9.6, then Θ is greater than 90° and Z is outside the operating zone of the relay, which then will not operate. Being nondirectional, the impedance relay will operate for all faults along the vector AB (see Figure 9.7) and for all faults behind the busbar, i.e. along the vector AC. The vector AB represents the impedance in front of the relay between its location at A and the end of the line AB, while the vector AC represents the impedance of the line behind the site of the relay.

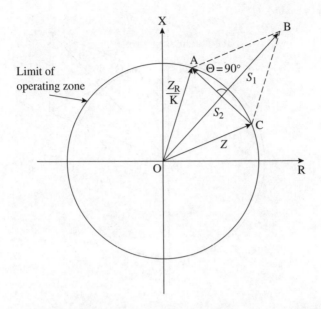

Figure 9.4   Operating characteristic of an impedance relay obtained using a phase comparator

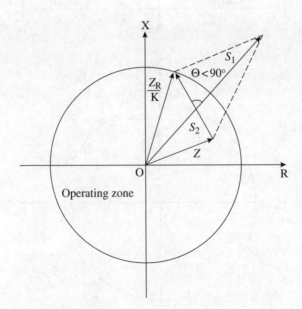

Figure 9.5   Impedance Z inside the operating zone of an impedance relay

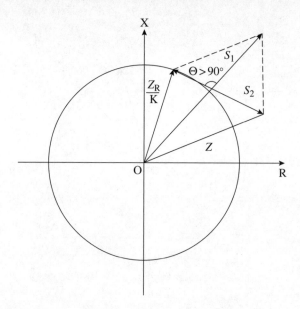

Figure 9.6   Impedance Z outside the operating zone of an impedance relay

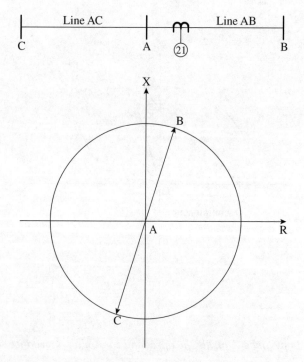

Figure 9.7   Impedance relay characteristic in the complex plane

The impedance relay has three main disadvantages:

1. It is not directional; it will see faults in front and behind its location and therefore requires a directional element in order to obtain correct discrimination. This can be obtained by adding an independent directional relay to restrict or prevent the tripping of the distance relay when power flows out of the protected zone during a fault.
2. It is affected by the arc resistance.
3. It is highly sensitive to oscillations on the power system, due to the large area covered by its circular characteristic.

9.2.2 Directional relay

Directional relays are elements that produce tripping when the impedance measured is situated in one half of the R-X plane. They are commonly used together with impedance relays in order to limit the operating zone to a semi-circle.

The operating characteristic is obtained from a phase comparison of the following signals:

$$S_1 = KV$$
$$S_2 = Z_R I \tag{9.9}$$

Dividing by KI, and defining $Z = V/I$, gives

$$S_1 = Z$$
$$S_2 = Z_R/K \tag{9.10}$$

The operating zone of the directional relay is defined by the values of Z and Z_R, which result in a phase difference between S_1 and S_2 of less than 90°. The construction of the characteristic is shown in Figure 9.8, in which S_1 and S_2 are drawn.

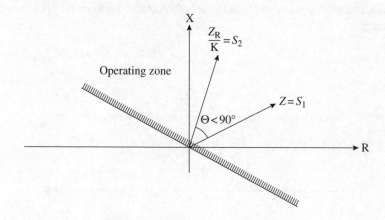

Figure 9.8   Operating zone of a directional relay

9.2.3   Reactance relay

The reactance relay is designed to measure only the reactive component of the line impedance; consequently, its setting is achieved by using a value determined by the reactance X_R. In this case the pair of equations for S_1 and S_2 is as follows:

$$S_1 = -KV + X_R I$$
$$S_2 = X_R I$$

(9.11)

and, dividing by KI, gives

$$S_1 = -Z + X_R/K$$
$$S_2 = X_R/K$$

(9.12)

The operating characteristics are obtained by drawing eqns. 9.12 in the complex plane and determining those values of Z for which Θ is less than $90°$. The construction is shown in Figure 9.9; here, the limit of the operating zone is a straight line parallel to the resistance axis, drawn for a reactance setting of X_R/K.

As the impedance of the fault is almost always resistive, it might be assumed that the fault resistance has no effect on the reactance relays. In a radial system this is generally true, but not necessarily if the fault is fed from two or more points since the voltage drop in the fault resistance is added to the drop in the line and affects the relay voltage. Unless the current in the relay is exactly in phase with the fault current, the voltage drop in the fault resistance will result in a component $90°$ out of phase to the relay current, producing an effect similar to the line reactance. This apparent reactance can be positive or negative and add to, or subtract from, the impedance measured by the relay, thus affecting its operation. If the fault resistance is large in comparison to the line reactance the effect could be serious and this type of relay should not be used.

Figure 9.10 shows the voltage seen by the relay in the presence of faults with arc resistance and two infeeds. From the diagram it will be seen that the relay will measure a value that is smaller than the actual reactance between the relay point and the fault.

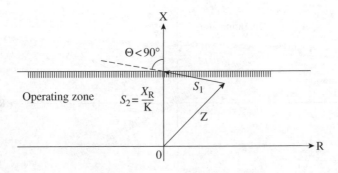

Figure 9.9   Operating zone of a reactance relay

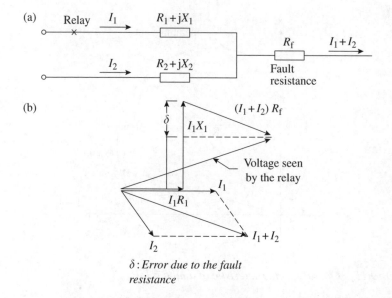

Figure 9.10   Voltage seen by a relay in the presence of faults with arc resistance and two infeeds: (a) *schematic of circuit;* (b) *vector diagram*

9.2.4   Mho relay

The mho relay combines the properties of impedance and directional relays. Its characteristic is inherently directional and the relay only operates for faults in front of the relay location; in addition it has the advantage that the reach of the relay varies with the fault angle. The characteristic, drawn in the R-X plane, is a circle with a circumference that passes through the origin of the co-ordinates and is obtained by assigning the signals the following values:

$$S_1 = -KV + Z_R I$$
$$S_2 = KV$$

(9.13)

from which

$$S_1 = -Z + Z_R/K$$
$$S_2 = Z$$

(9.14)

Drawing Z_R/K and the eqns. 9.14 in the R-X plane, the relay characteristic is determined by the locus for the values of Z that are fulfilled when Θ is less than 90°. In this case the limit of the operating zone ($\Theta = 90°$), as shown in Figure 9.11, is traced by a circle with a diameter of Z_R/K and a circumference that passes through the origin of the co-ordinates. For values of Z located inside the circumference, Θ will be less than 90°, as is shown in Figure 9.12, and this will result in operation of the relay.

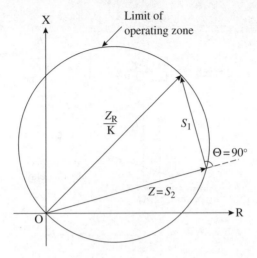

Figure 9.11   Operating characteristic of a mho relay

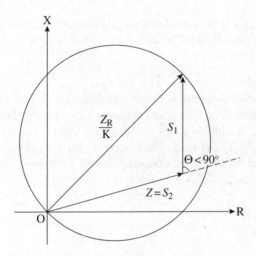

Figure 9.12   Impedance Z inside the operating zone of a mho relay

9.2.5   *Completely polarised mho relay*

One of the disadvantages of the autopolarised mho relay is that, when it is used on long lines and the reach does not cover the section sufficiently along the resistance axis, then it is incapable of detecting faults with high arc or fault resistances. The problem is aggravated in the case of short lines since the setting is low and the amount of the R axis covered by the mho circle is small in relation to the values of arc resistance expected.

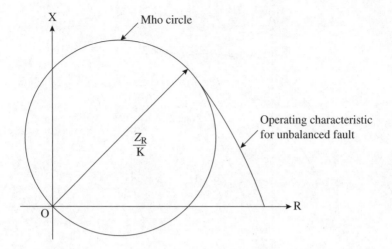

Figure 9.13   Operating characteristic of a completely polarised mho relay

One practical solution to this problem is to use a completely polarised mho relay where the circular characteristic is extended along the R axis for all unbalanced faults, as illustrated in Figure 9.13. This characteristic can be obtained by means of a phase comparator, which is fed by the following signals:

$$S_1 = V_{pol}$$
$$S_2 = V - IZ_R \tag{9.15}$$

where: V = voltage at the location of the relay, on the faulted phase or phases; V_{pol} = polarisation voltage taken from the phase, or phases, not involved with the fault; I = fault current; Z_R = setting of the distance relay.

9.2.6   Relays with lens characteristics

Distance relays with lens characteristics are very useful for protecting high impedance lines that have high power transfers. Under these conditions the line impedance values, which are equal to V^2/S, become small and get close to the impedance characteristics of the relay, especially that of zone 3. This offset lens characteristic, which can be adjusted to offset the mho circular characteristic as shown in Figure 9.14, is common in some relays.

9.2.7   Relays with polygonal characteristics

Relays with polygonal characteristics provide an extended reach in order to cover the fault resistance, in particular for short lines, since the position of the resistance line can be set in the tripping characteristic (see line 2 in Figure 9.15 which shows a typical polygonal operating characteristic).

The polygonal tripping characteristic is obtained from three independent measuring elements – reactance, resistance and directional. In order to achieve this

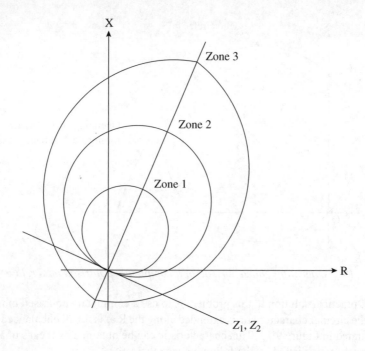

Figure 9.14   Zone 3 offset lens characteristic

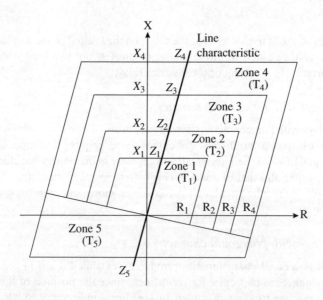

Figure 9.15   Polygonal relay operating characteristic

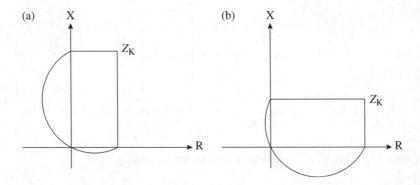

Figure 9.16 *Typical combined operating characteristics:* (a) R/X ratio = 0.5;
(b) R/X ratio = 2

characteristic, the measuring elements are suitably combined. The relay is tripped
only when all three elements have operated; in this way the required polygonal
characteristic is obtained.

9.2.8 Relays with combined characteristics

A typical combined operating characteristic is defined in the impedance plane by lines
running parallel to the resistive and reactive axes which cross each other at the setting
point for Z_K, as shown in Figure 9.16. In order to achieve the required directionality,
a mho circle that passes through Z_K is employed. In relays with this characteristic the
reaches in the resistive and reactive directions have the same range of settings and
can be adjusted independently of each other.

9.3 Setting the reach and operating time of distance relays

Distance relays are set on the basis of the positive-sequence impedance from the relay
location up to the point on the line to be protected. Line impedances are proportional
to the line lengths and it is this property that is used to determine the position of the
fault, starting from the location of the relay. However, this value is obtained by using
system currents and voltages from the measurement transformers that feed the relays.
Therefore, in order to convert the primary impedance into a secondary value, which
is used for the setting of the distance relay, the following expression is used:

$$\frac{V_{prim}}{I_{prim}} = Z_{prim} = \frac{V_{sec} \times VTR}{I_{sec} \times CTR} \tag{9.16}$$

Thus

$$Z_{sec} = Z_{prim} \times (CTR/VTR) \tag{9.17}$$

where CTR and VTR are the transformation ratios of the current and voltage
transformers, respectively.

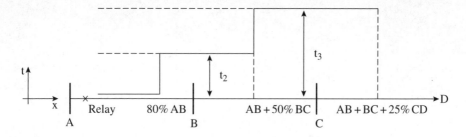

Figure 9.17   Distance relay protection zones for a radial system

Normally, three protection zones in the direction of the fault are used in order to cover a section of line and to provide back-up protection to remote sections (see Figure 9.17). Some relays have one or two additional zones in the direction of the fault plus another in the opposite sense, the latter acting as a back-up to protect the busbars. In the majority of cases the setting of the reach of the three main protection zones is made in accordance with the following criteria:

- zone 1: this is set to cover between 80 and 85 per cent of the length of the protected line;
- zone 2: this is set to cover all the protected line plus 50 per cent of the shortest next line;
- zone 3: this is set to cover all the protected line plus 100 per cent of the second longest line, plus 25 per cent of the shortest next line.

In addition to the unit for setting the reach, each zone unit has a timer unit. The operating time for zone 1, t_1, is normally set by the manufacturer to trip instantaneously since any fault on the protected line detected by the zone 1 unit should be cleared immediately without the need to wait for any other device to operate. The operating time for zone 2 is usually of the order of 0.25 to 0.4 s, and that of zone 3 is in the range of 0.6 to 1.0 s. When there are power transformers at adjacent substations the zone 2 timer should have a margin of 0.2 s over the tripping time of any associated transformer overcurrent protection. In the case of zone 3, when the settings of relays at different locations overlap, then the timer for the zone 3 of the furthest relay should be increased by at least 0.2 s to avoid incorrect co-ordination. However, the operating time for the zone 3 units should also be set at a value that will ensure that system stability is maintained and therefore, if necessary, consideration may have to be given to reducing the zone 3 operating time in such circumstances.

Since the tripping produced by zone 1 is instantaneous, it should not reach as far as the busbar at the end of the first line (see Figure 9.17) so it is set to cover only 80–85 per cent of the protected line. The remaining 20–15 per cent provides a factor of safety in order to mitigate against errors introduced by the measurement transformers and line impedance calculations. The 20–15 per cent to the end of the line is protected by zone 2, which operates in t_2 s. Zone 3 provides the back-up and operates with a delay of t_3 s. Since the reach and therefore the operating time of the

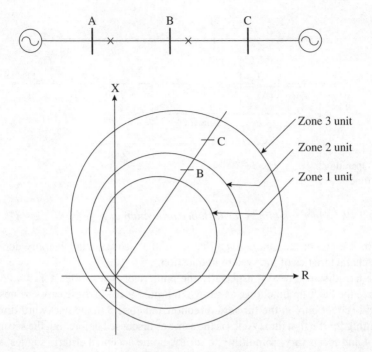

Figure 9.18   Operating characteristic of distance protection located at A

distance relays are fixed, their co-ordination is much easier than that for overcurrent relays.

In order to illustrate the philosophy referred to earlier, consider the case of the system in Figure 9.18 in which it is required to protect the lines AB and BC. For this, it is necessary to have three relays at A to set the three zones. All three units should operate for a fault within the operating characteristic of zone 1. For a fault on line BC, but within the cover of the zone 2 unit at A, both the zone 2 and zone 3 units should operate. Since there is also distance protection at substation B, the relay at A should provide an opportunity for the breaker at B to clear the fault; it is for this reason that the zone 2 and zone 3 units operate with an appropriate time delay in order to obtain discrimination between faults on lines AB and BC. The diagram of operating times is shown in Figure 9.19.

Some methods for setting distance relays use different criteria to those already mentioned, mainly with regard to the reach of zones 2 and 3. In particular, there is the method where it is recommended that the reach of zone 2 should be 120 per cent of the impedance of the line to be protected, and that for zone 3 should be 120 per cent of the sum of impedances of the protected line and of its longest adjacent line. In this case the times for zones 2 and 3 should not have a fixed value, but should be based on the opening time of the breakers and the reach of the relays to guarantee that there will be no overlap in the same zones covered by adjacent relays. Since the same philosophy is used as the basis for either method, no specific recommendation is

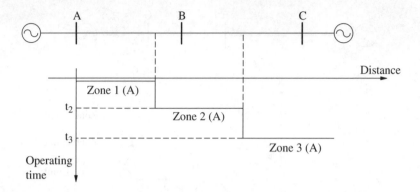

Figure 9.19   Operating times for distance protection at A

made to use one or the other, given that the actual selection is generally dependent on the characteristics of the system in question.

Modern distance relays, especially the numerical types, offer zones 4 and 5 to reinforce the back-up functions, as shown in Figure 9.15. In these cases, zones 3 and 4 provide cover only in the forward direction and zone 5 in the backward direction. The setting for the first three zones is the same as discussed before, but the settings for zones 4 and 5 can vary from utility to utility. Some accepted criteria suggest setting zone 4 at 120 per cent of zone 3, and zone 5 at 20 per cent of zone 1. The time delay for zones 4 and 5 is normally the same as that for zone 3 but increased by a margin of, typically, 400 ms. Care should be taken to ensure that the zones with the higher settings, i.e. zones 3 and 4, do not overlap different voltage levels through step-up or step-down transformers, or load impedance values.

9.4   The effect of infeeds on distance relays

The effect of infeeds needs to be taken into account when there are one or more generation sources within the protection zone of a distance relay which can contribute to the fault current without being seen by the distance relay.

Analysing the case illustrated in Figure 9.20, it can be appreciated that the impedance seen by the distance relay at A for a fault beyond busbar B is greater than actually occurs. In fact, if a solid earth-fault is present at F, the voltage at the relay at A would be

$$V_A = I_A Z_A + (I_A + I_B) Z_B \tag{9.18}$$

from which

$$\frac{V_A}{I_A} = Z_A + \left[1 + \frac{I_B}{I_A}\right] Z_B \tag{9.19}$$

The relay therefore sees an impedance of KZ_B, the infeed constant K being equal to I_B/I_A, in addition to the line impedance Z_A, which implies that its reach is reduced.

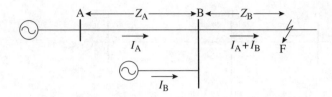

Figure 9.20   Effect of an infeed on distance protection

The setting of zones 2 and 3 for the relay at A should then take the following form:

$$Z_{relay} = Z_A + (1 + K)Z_B \qquad (9.20)$$

where K is given as

$$K = \frac{I_{total\ infeed}}{I_{relay}} \qquad (9.21)$$

It is necessary to take into account the fact that the distance relay can over reach if the sources that represent the infeed are disconnected, so that a check should be made for these conditions to ensure that there is no overlap with the adjacent zone 2. For systems in which zones 2 and 3 cover lines that are not part of a ring, the value of K is constant and independent of the location of the fault, given the linearity of electrical systems. By way of illustration, consider the system shown in Figure 9.21a. Figure 9.21b shows the impedance that is seen by a distance relay located in substation C. For a fault between B and D, the value of K will be the same for faults at either substation B or D, or at some other point on the line linking the two substations. If the fault location moves from B to D, the current values diminish but the ratio of the total infeed to the current seen by the relay will be the same.

Since the value of the infeed constant depends on the zone under consideration, several infeed constants, referred to as K_1, K_2 and K_3, need to be calculated. K_1 is used to calculate the infeed for the second zone. K_2 and K_3 are used for zone 3, K_2 taking account of the infeed on the adjacent line and K_3 that in the remote line. Based on the criteria in Section 9.3, and considering the infeed as discussed before, the expressions for calculating the reach of the three zones for the relay located at busbar C would be

$$Z_1 = 0.8 \text{ to } 0.85 \text{ times } Z_{AB}$$

$$Z_2 = Z_{CB} + 0.5(1 + K_1)Z_{BD}$$

$$Z_3 = Z_{CB} + (1 + K_2)Z_{BF} + 0.25(1 + K_3)Z_{FH}$$

where

$$K_1 = \frac{I_A + I_E + I_F}{I_C}$$

$$K_2 = \frac{I_A + I_D + I_E}{I_C}$$

$$K_3 = \frac{I_A + I_D + I_E + I_G}{I_C}$$

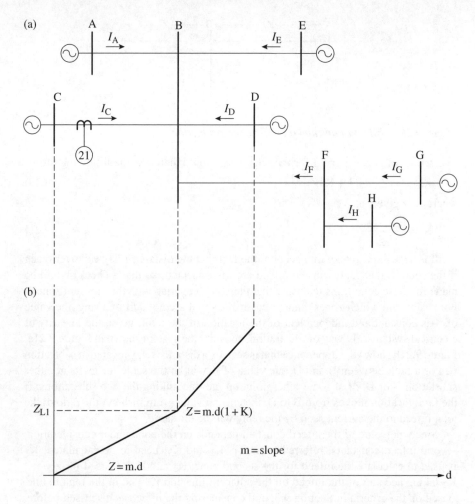

Figure 9.21   *Distance protection with a fixed value of K:* (a) typical system for analysis of infeed with fixed K; (b) impedance seen by distance relay at C

Since the coverage on the remote line is not very critical, K_3 can be assumed to be equal to K_2, which reduces the reach of the zone 3. However, this difference is normally negligible, and the assumption is therefore acceptable.

In drawing the impedance as a function of the distance, as shown in Figure 9.21b, it can be seen that the slope of line CB is constant, as is to be expected. However, in section BD the slope continues to be constant, although of a different value to that in the section CB, due to the inclusion of the term K. The important point is that the impedance seen by the relay at C for faults in section BD is directly proportional to the location of the fault, the same as for faults in section CB. When zones 2 and 3 cover part of the lines of a double circuit, or of rings, the value of K will depend on

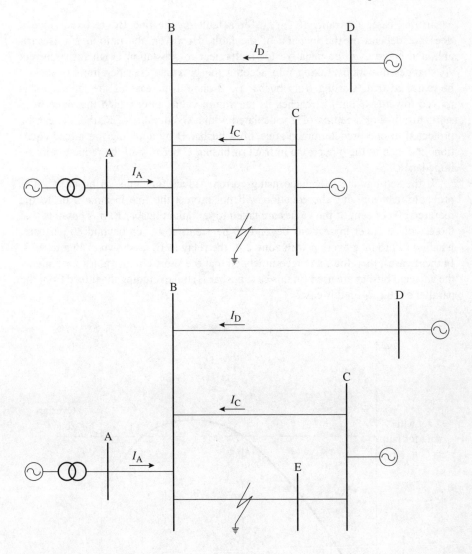

Figure 9.22   Distance protection with variable K

the fault point or busbar under consideration, as is evident from the two schematics shown in Figure 9.22. In both cases the impedance that the relay sees for faults along the line BC is a curve of varying slope.

If the infeed equation is split, then the impedance that the relay sees for a fault in section BC of the two schemes in Figure 9.22 can be written as

$$Z_{\text{relay}} = Z_{\text{line}} + \left[1 + \frac{I_D + I_C}{I_A} \right] Z_B \text{ to fault} \tag{9.22}$$

In this case, the ratio of I_D/I_A for a fault on the line BC is fixed since it does not depend on the location of the fault. However, the ratio I_C/I_A is variable and could even be negative for faults near to substation C when the current I_C changes direction. Taking into account the previous considerations, it should be expected that, starting from busbar B, Z should increase as the fault point is moved towards C until it reaches its maximum value, after which the value of Z starts to fall. For a setting of Z calculated with a fixed value of K, the line can be protected in one predetermined zone from busbar B by applying the infeed equation, and then in the other zone in front of busbar C because of the reduction in the impedance.

If the settings are calculated using current values for faults on busbar C, it is probable that applying the equation will not protect the line in zone 2 up to the required 50 per cent. If the values are taken for a fault at busbar B, it is possible that there will be cover in sections beyond 50 per cent, and even beyond 80 per cent, which will produce overlap with zone 2 of the relay at B, as shown in Figure 9.23. In these cases, therefore, it is recommended that the short-circuit values for faults at the adjacent busbar are used (in this case busbar B) but excluding the infeed from the parallel circuit, I_C in this case.

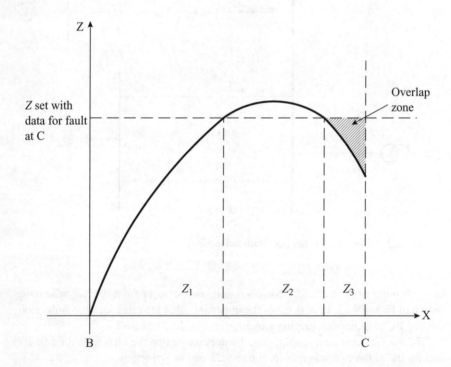

Figure 9.23   *Overlap in a ring system. $Z_1 =$ zone protected by zone 2; should be greater than 50 per cent of BC. $Z_2 =$ unprotected zone. Z_3 represents overlap zone seen by the zones 2 of the distance relays at A and B*

9.5 The effect of arc resistance on distance protection

For a solid fault, the impedance measured by the relay is equal to the impedance up to the fault point. However, the fault may not be solid, i.e. it might involve an electric arc or an impedance. With arc faults it has been found that the voltage drop in the fault and the resultant current are in phase, indicating that the impedance is purely resistive. When dealing with an earth fault, the fault resistance is made up of the arc resistance and the earth resistance. Faults with arc resistance are critical when they are located close to the limits of the relay protection zones since, although the line impedance is inside the operating characteristic, the arc resistance can take the total resistance seen by the relay outside this characteristic resulting in under reaching in the relay. The situation for an impedance type relay, where the effect is particularly critical, is given in Figure 9.24.

It can be seen from Figure 9.25 that, if the characteristic angle of the relay, ϕ, has been adjusted to be equal to the characteristic angle of the line Θ then, under fault conditions with arcing present, the relay will under reach. For this reason it is common practice to set ϕ a little behind Θ (by approximately ten degrees) in order to be able to accept a small amount of arc resistance without producing under reaching.

From Figure 9.25, and taking into account that an angle inscribed inside a semicircle is a right angle:

$$Z = (Z_R/K) \cos(\Theta - \phi) \tag{9.23}$$

If the angle of the protected line is equal to that of the relay, the settings are correct. However, if the angle of the line exceeds that of the relay by ten degrees, the

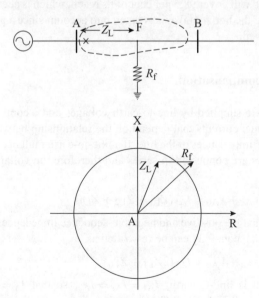

Figure 9.24   Under reach of an impedance relay due to arc resistance

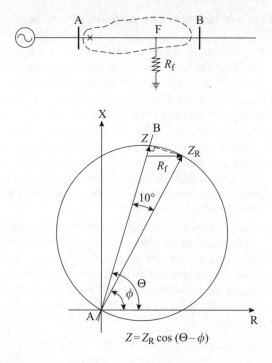

Figure 9.25   Mho relay setting for arc faults

relay characteristic will cover 98.5 per cent of its reach, which is acceptable when the higher cover along the horizontal axis is taken into account since a greater resistance coverage is achieved.

9.6   Residual compensation

Earth-fault units are supplied by line-to-earth voltages and a combination of phase currents and residual currents that depend on the relationship between the positive- and zero-sequence impedances of the line. If a line-to-earth fault occurs, say A-E, the sequence networks are connected in series and therefore the voltage applied to the relay is

$$V_A = V_{A1} + V_{A2} + V_{A0} = I_{A1}(Z_{L1} + Z_{L2} + Z_{L0}) \tag{9.24}$$

and, on the basis that the positive and negative-sequence impedances of a line can be assumed to be equal, eqn. 9.24 can be re-written as

$$V_A = V_{A1} + V_{A2} + V_{A0} = I_{A1}(2Z_{L1} + Z_{L0}) \tag{9.25}$$

Since the fault is line-to-earth, $I_{A1} = I_{A2} = I_{A0}$, so that $I_A = 3I_{A1}$. The ratio V_A/I_A is therefore $(2Z_{L1} + Z_{L0})/3$ which does not equal the positive-sequence impedance Z_{L1}.

The value of residual current to be injected is calculated so that a relay that is set to the positive-sequence impedance of the line operates correctly. Therefore, applying the line and residual currents to the relay

$$I_A + 3K I_{A0} = I_A (1 + K) \tag{9.26}$$

and

$$\frac{V_A}{I_A (1 + K)} = Z_{L1} \tag{9.27}$$

Replacing V_A / I_A gives

$$\frac{2 Z_{L1} + Z_{L0}}{3} = Z_{L1} (1 + K) \tag{9.28}$$

from which

$$K = \frac{Z_{L0} - Z_{L1}}{3 Z_{L1}} \tag{9.29}$$

9.7 Impedances seen by distance relays

Distance relays are designed to protect power systems against four basic types of fault – three-phase, phase-to-phase, phase-to-phase-to-earth, and single-phase faults. In order to detect any of the above faults, each one of the zones of distance relays requires six units – three units for detecting faults between phases (A-B, B-C, C-A) and three units for detecting phase-to-earth faults (A-E, B-E, C-E). A complete scheme would have one set of these six units for each zone, although so-called switching schemes use one set for one or more zones. The setting of distance relays is always calculated on the basis of the positive-sequence impedance. Given the impossibility of selecting exactly the right voltages and currents to cover all types of faults, each unit receives a supply that is independent of the others in order to obtain the required relay operation.

9.7.1 Phase units

Phase units are connected in delta and consequently receive line-to-line voltages and the difference of the line currents. The impedances that they measure are a ratio of voltages and currents as follows:

$$\begin{aligned} Z_{AB} &= \frac{V_{AB}}{I_A - I_B} \\[1em] Z_{BC} &= \frac{V_{BC}}{I_B - I_C} \\[1em] Z_{CA} &= \frac{V_{CA}}{I_C - I_A} \end{aligned} \tag{9.30}$$

9.7.2   *Earth-fault units*

As mentioned earlier in Section 9.5, earth-fault units are supplied by line-to-earth voltages and a combination of phase and residual currents. From the previous calculations, the impedances measured by the earth-fault units of distance relays for the three phases are

$$Z_A = \frac{V_A}{I_A\left[1 + \dfrac{Z_0 - Z_1}{3Z_1}\right]} = \frac{V_A}{I_A + \dfrac{Z_0 - Z_1}{Z_1}I_0}$$

$$Z_B = \frac{V_B}{I_B\left[1 + \dfrac{Z_0 - Z_1}{3Z_1}\right]} = \frac{V_B}{I_B + \dfrac{Z_0 - Z_1}{Z_1}I_0} \tag{9.31}$$

$$Z_C = \frac{V_C}{I_C\left[1 + \dfrac{Z_0 - Z_1}{3Z_1}\right]} = \frac{V_C}{I_C + \dfrac{Z_0 - Z_1}{Z_1}I_0}$$

9.8   Power system oscillations

Power system oscillations can occur, for example, after a short-circuit has been removed from the system, or when switching operations are carried out that involve the connection or disconnection of large quantities of load. During this phenomenon the voltage and current that feed the relay vary with time and, as a result, the relay will also see an impedance that is varying with time, which may cause it to operate incorrectly.

To illustrate the situation involving a distance relay during such oscillations, consider the equivalent circuit of the power system shown in Figure 9.26. Assume that there is a transfer of power from the source of supply, S, to the most distant load at R. The current, I_S, which flows from S towards R causes a voltage drop in the system elements in accordance with the vector diagram shown in Figure 9.27. The value of Θ_S, the phase difference between E_S and E_R, increases with the load transferred.

The impedance measured by the distance relay situated at A is $Z = V_A/I_S$; the expression for this impedance can be obtained starting from the voltage V_A which

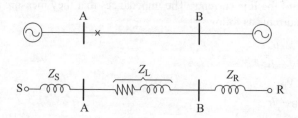

Figure 9.26   Equivalent circuit for analysis of power system oscillations

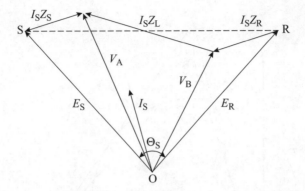

Figure 9.27   Vector diagram for power system oscillation conditions

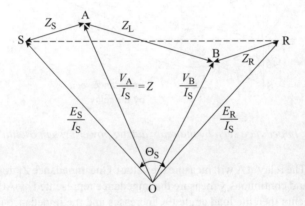

Figure 9.28   Impedance diagram for system in Figure 9.26

supplies the relay:

$$V_A = I_S Z_L + I_S Z_R + E_R$$
$$V_A / I_S = Z_L + Z_R + E_R / I_S$$
(9.32)

The last equation can be easily drawn by dividing the vectors in Figure 9.27 by the oscillation current I_S. In this way the diagram of system impedances, which is shown in Figure 9.28, is obtained in which all the parameters can be assumed to be constant except I_S and Θ_S, which are variable and depend on the power transfer. The increment of load transferred brings with it an increase in I_S and Θ_S. This results in a reduction in the size of the vector V_A / I_S (see Figure 9.28) and, if the increment of load is sufficiently large, the impedance seen by the relay (V_A / I_S) can move into the relay operating zones, as shown in Figure 9.29.

Figure 9.29 is obtained by constructing an R-X plane over the locus of the relay A, and then drawing over this the relay operating characteristic and the diagram of system

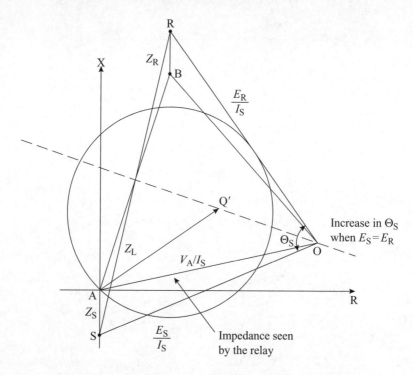

Figure 9.29   Impedance seen by the relay during power system oscillations

impedances. The relay at A will measure the value of the impedance Z_L for a solid fault to earth at B and continuously measure the impedance represented by AO. If a severe oscillation occurs then the load angle Θ_S increases and the impedance measured by the relay will decrease to the value AQ′, which can be inside the relay operating characteristic. The locus of the impedance seen by the relay during oscillations is a straight line when $E_S = E_R$, as in Figure 9.29. If $E_S > E_R$, the locus is a family of circles centred on the SR axis. A typical trajectory which delineates the impedance in the R-X plane during a power oscillation is shown in Figure 9.30. Consequently, the trajectory passes inside the relay operating characteristic, indicating that there will be a possibility for the associated breaker to be tripped in the presence of system oscillations.

In order to prevent the operation of the relay during oscillations, a blocking characteristic is used (see Figure 9.30). The trajectory of the impedance crosses the characteristics of the measuring and blocking units. If the measuring units operate within a given time, and after the blocking unit has operated, tripping of the breaker is permitted. On the other hand, if the measuring units have not operated after a predetermined time delay, the breaker will not be tripped. Thus, under fault conditions when the blocking and measuring units operate virtually simultaneously, tripping takes place. However, under power oscillation conditions, when the measuring units operate some time after the blocking unit, tripping is prevented.

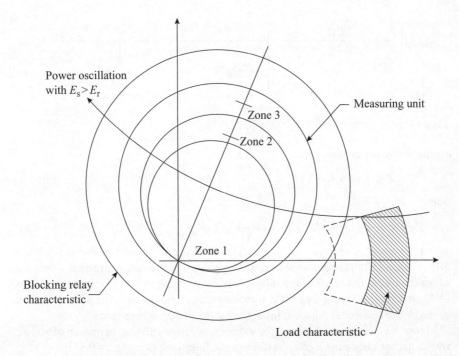

Figure 9.30   Blocking characteristic to prevent relay operation during power system oscillations

To prevent operation of the relay during oscillations, a power-swing blocking unit is added. The diameter, or reach, of its characteristic for mho relays is generally 1.3 or more times the diameter of the outermost zone of the relay, which is usually zone 3. During fault conditions the displacement of the impedance value seen by a distance relay is much faster than during power swings. This fact is used to set the power swing blocking unit, which is then inhibited if there is a time elapse of typically 0.1 s or less, to enable the impedance trajectory to move from the power-swing blocking characteristic into zone 3 or outermost relay characteristic. Manufacturers will usually supply recommendations for setting this unit, when provided, depending on the actual relay types being used, and the values given above should therefore be used as general guidelines only.

9.9   The effective cover of distance relays

In interconnected power systems in which there are supply infeeds, the effective reach of distance relays does not necessarily correspond to the setting value in ohms. It is possible to calculate the ratio between both, using infeed constants that have been defined earlier. The setting value of distance relays for zones 2 and 3 is determined

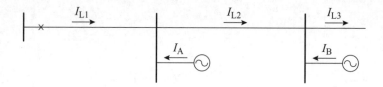

Figure 9.31   Power system with multiple infeeds

by the following expressions:

$$Z_2 = Z_{L1} + (1 + K_1)X_2 Z_{L2} \tag{9.33}$$

and

$$Z_3 = Z_{L1} + (1 + K_2)X_2 Z_{L2} + (1 + K_3)X_3 Z_{L3} \tag{9.34}$$

where X_2 and X_3, the percentage of effective cover as defined in Section 9.3, are 50 per cent and 25 per cent respectively. However, in some cases, principally because of limitations in the reach of the relays, it is not possible to set the calculated values of Z_2 and Z_3 in the relay, and it is therefore necessary to assess the effective cover given by the relay over adjacent lines against the actual setting value.

From the previous equations, the expression for calculating the cover of zone 2 over adjacent lines can be calculated from the following (see Figure 9.31).

$$X_2 = \frac{Z_2 - Z_{L1}}{Z_{L2}(1 + K_1)} \tag{9.35}$$

where: Z_2 = setting for zone 2 in ohms; $X_2 Z_{L2}$ = effective cover over adjacent line in ohms; Z_{L1} = impedance of line associated with relay; K_1 = infeed constant for the adjacent line.

From the above equations it can be found that the expression for calculating the effective cover of the relay over remote lines is

$$X_3 = \frac{Z_3 - Z_{L1} - (1 + K_2)Z_{L2}}{Z_{L3}(1 + K_3)} \tag{9.36}$$

where: Z_3 = zone 3 setting in ohms; $X_3 Z_{L3}$ = effective cover over remote line in ohms; Z_{L1} = impedance of line associated with relay; K_2 = infeed constant for the longest adjacent line; Z_{L2} = impedance of the longest adjacent line; K_3 = infeed constant for the remote line.

Using these equations and the appropriate infeed constants, it is possible to calculate the effective reach of the relay over any of the lines adjacent to the protected line.

9.10   Maximum load check

This check is made to ensure that the maximum load impedance will never be inside the outermost characteristic, which is normally zone 3. To fulfil this, the distance

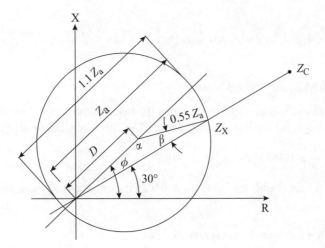

Figure 9.32   Check of maximum load for mho relay

between the characteristic of zone 3 and the maximum load point should be at least 25 per cent of the distance between the origin and the maximum load point for single circuit lines, and 50 per cent for double circuit lines.

Mho relays

Typically the mho relay operating characteristic for zone 3 has a displacement (offset) of 10 per cent of the setting value, as illustrated in Figure 9.32. The maximum load point is defined as

$$Z_c = \frac{V^2}{S_{max}} \angle 30° \tag{9.37}$$

In the diagram, Z_a and ϕ are the setting and the characteristic angle of the relay, respectively. From Figure 9.32

$$D = Z_a - \frac{1.1 \, Z_a}{2} = 0.45 Z_a \tag{9.38}$$

Applying the sine theory

$$\sin \beta / \sin(\phi - 30°) = (0.45 Z_a)/(0.55 Z_a) \tag{9.39}$$

from which

$$\sin \beta = 0.818 \sin(\phi - 30°) \tag{9.40}$$

From the previous expression the value of α can be obtained. Also, from Figure 9.32

$$\alpha = 180° - \beta - (\phi - 30°)$$

and

$$\sin\alpha/\sin(\phi-30°)=Z_x/(0.55Z_a) \tag{9.41}$$

Therefore:

$$Z_x=0.55Z_a\sin\alpha/\sin(\phi-30°)$$

For all cases, it is possible to calculate the reach of the relay in the direction of the load by applying the last equation above. The check consists of verifying that

$$\frac{Z_c-Z_x}{Z_c}\times 100\%\geq P \tag{9.42}$$

where $P=0.5$ for double circuit lines, and 0.25 for single circuit lines, as mentioned earlier.

Relays with a polygonal characteristic

Zone 3 will be determined by the reactive and resistive settings, X_3 and R_3, respectively. The situation is shown in Figure 9.33.

From Figure 9.33 it can be seen that:

$$\phi=\tan^{-1}(X_3/R_3),\quad r=\sqrt{R_3^2+X_3^2}$$

$$(\phi-30°)+120°+\beta=180° \tag{9.43}$$

$$\beta=90°-\phi$$

Applying the sine theory

$$\frac{\sin\beta}{\sin120°}=\frac{Z_x}{r}$$

$$Z_x=r\frac{\sin\beta}{\sin120°} \tag{9.44}$$

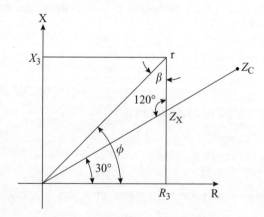

Figure 9.33   Check of maximum load for polygonal relay

The above equation enables the reach of polygonal relays in the direction of the load to be determined. The distance Z_x should satisfy the condition defined by the inequality check, given above.

9.11 Drawing relay settings

The setting of distance relays can be represented in diagrams as time against reach in ohms for the item of equipment being protected. The reach clearly depends on the settings having been defined in accordance with the methodology set out in the previous paragraphs. It should be noted that the settings calculated using the equations are subject to two restrictions:

1. Limitations for the particular relay, when the calculated value is excessively high and it is impossible to set the relay.
2. Limitations for the load, when the value for the reach of zone 3 approaches the maximum load point too closely.

When the first type of restriction applies, the reach is adjusted to the maximum available on the relay.

Example 9.1

The following case study illustrates the procedure that should be followed to obtain the settings of a distance relay. Determining the settings is a well-defined process, provided that the criteria are correctly applied, but the actual implementation will vary, depending not only on each relay manufacturer but also on each type of relay.

For the case study, consider a distance relay installed at the Pance substation in the circuit to Juanchito substation in the system shown diagrammatically in Figure 9.34, which provides a schematic diagram of the impedances as seen by the relay. The numbers against each busbar correspond to those used in the short-circuit study, and shown in Figure 9.35. The CT and VT transformation ratios are 600/5 and 1000/1 respectively.

From the criterion for setting zone 1

$$Z_1 = 0.85Z_{10-11} = 0.85(7.21\angle 80.5°) = 6.13\angle 80.5° \text{ primary ohms}$$

and for zone 2

$$Z_2 = Z_{10-11} + 0.5(1 + K_1)Z_{11-9}$$

In this case the infeed constant is defined as:

$$K_1 = \frac{I_{14-11} + I_{17-11} + I_{5-11} + I_{18-11}}{I_{10-11}}$$

for a fault at busbar 11.

From the short-circuit values in Figure 9.35

$$K_1 = \frac{1333.8\angle -85.54° + 0 + 5364.6\angle -85.88° + 449.9\angle -86.34°}{2112.6\angle(-85.55°)}$$

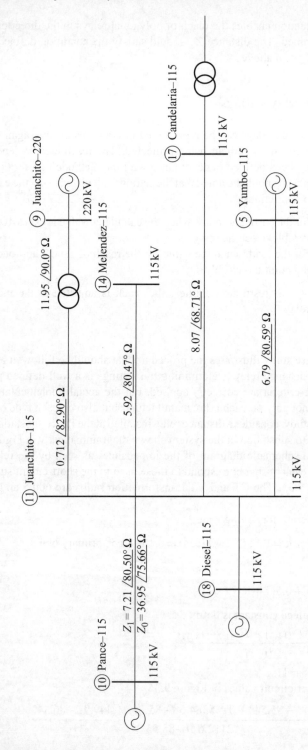

Figure 9.34   Impedance diagram for Example 9.1 showing impedances seen by a relay on the Juanchito circuit at Pance substation

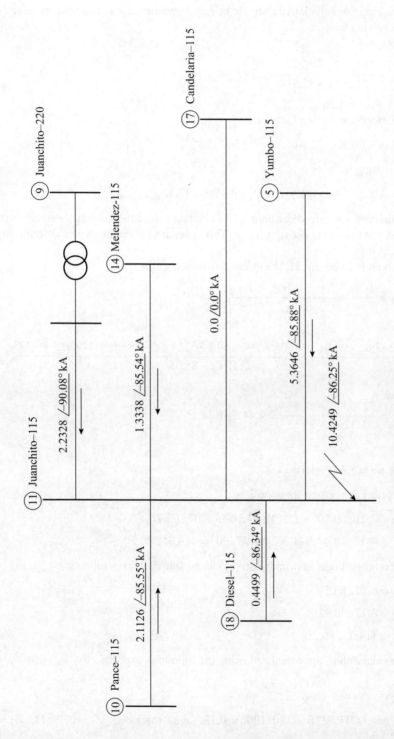

Figure 9.35 Fault current contribution for Example 9.1

In making this calculation it should be noted that the values of current are referred to the receiving busbar.

This gives

$$K_1 = \frac{7148.27\angle-85.87°}{2112.6\angle-85.5°} = 3.38\angle-0.37°$$

so that $1+K_1 = 4.38$.

Therefore, the setting for zone 2 is

$$Z_2 = 7.21\angle80.50° + (4.38 \times 0.356\angle82.90°) = 8.77\angle80.93° \text{ primary ohms}$$

and the setting for zone 3 is

$$Z_3 = Z_{10-11} + (1+K_2)Z_{11-17} + 0.25(1+K_3)Z_{\text{transformer}}$$

In this case, the infeed constant K_3 will be taken to be the same as K_2 since under reach on that section is not significant. This approach is common when determining zone 3 settings.

For a fault on busbar 11, the infeed constant is defined as

$$K_2 = \frac{I_{9-11} + I_{14-11} + I_{5-11} + I_{18-11}}{I_{10-11}}$$

Thus:

$$K_2 = \frac{2232.8\angle-90.08° + 1333.8\angle-85.54° + 5364.6\angle-85.88° + 449.9\angle-86.34°}{2112.6\angle-85.55°}$$

i.e.

$$K_2 = \frac{9376.72\angle-86.86°}{2112.6\angle-85.5°} = 4.44\angle-1.36°$$

so that

$$1+K_2 = 5.44\angle-1.10°$$

Therefore the setting for zone 3 is

$$Z_3 = 7.21\angle80.50° + (5.44\angle-1.10° \times 8.07\angle68.71°)$$
$$+ (1+4.44\angle-1.36°)11.95\angle90° = 114.35\angle80.20°$$

The relay settings, in primary ohms, can be summed up as follows:

$$Z_1 = 6.13\angle80.5°$$
$$Z_2 = 8.77\angle80.93°$$
$$Z_3 = 114.35\angle80.20°$$

The secondary ohms are calculated using the following expression:

$$Z_{\text{sec}} = Z_{\text{prim}} \times \frac{\text{CTR}}{\text{VTR}}$$

In this case CTR/VTR $= 120/1000 = 0.12$, and, therefore, $Z_1 = 0.736\,\Omega$, $Z_2 = 1.052\,\Omega$, and $Z_3 = 13.72\,\Omega$.

Starting unit settings

The starting unit is set by taking 50 per cent of the maximum load impedance. Given that the power transferred from Pance to Juanchito is $S = 30.4 + j13.2\,\text{MVA} = 33.14\,\text{MVA}$, then

$$Z_c = \frac{V^2}{P} = \frac{115^2}{33.14} = 399.03 \text{ primary ohms} = 47.88 \text{ secondary ohms}$$

Residual compensation constant setting

$$K_1 = \frac{Z_0 - Z_1}{3Z_1}$$

where K_1 = residual compensation constant and Z_1, Z_0 = line positive- and zero-sequence impedances giving

$$K_1 = \frac{36.95\angle 75.66° - 7.21\angle 80.50°}{3(7.21\angle 80.50°)} = 1.377\angle -6°$$

$$K = 1.4$$

Time setting

Time delay for zone $2 = 0.4\,\text{s}$
Time delay for zone $3 = 1.0\,\text{s}$

Load check

The setting of the unit that determines the longest operating characteristic of the relay should be checked to make sure that it does not overlap the load zone.

The reach of the relay in the direction of the load is determined as follows:

$$\sin \beta = 0.818 \sin(\phi - 30°)$$

where the relay setting $\phi = 75°$.

$$\sin \beta = 0.818 \sin(45°)$$

i.e.

$$\beta = 35.34°$$

$$\alpha = 180° - \beta - (\phi - 30°)$$

giving $\alpha = 99.66°$ so that the reach will be

$$Z_x = \frac{0.55\, Z_3 \sin 99.66°}{\sin 45°} = \frac{0.55(114.35) \sin 99.66°}{\sin 45°}$$

$$Z_x = 87.68 \text{ primary ohms}$$

The distance to the load point, expressed as a percentage, is

$$\% = \frac{399.03 - 87.68}{399.03} \times 100\% = 78.03\%$$

Therefore, it is concluded that the setting is appropriate and does not require adjusting in reach as a result of the load.

Determination of the effective cover

In accordance with the calculated settings, in zone 2 the relay covers 50 per cent of the line 11-9. However, it is important to determine the cover of this setting along the Juanchito-Yumbo 115 kV line (11-5) and, for this, the following expression is used:

$$X_2 = \frac{Z_2 - Z_{L1}}{Z_{L2}(1 + K_1)}$$

The infeed constant K_1 at substation Juanchito (no. 11) is given by

$$K_1 = \frac{I_{9\text{-}11} + I_{14\text{-}11} + I_{17\text{-}11} + I_{18\text{-}11}}{I_{10\text{-}11}}$$

for a fault on busbar 11.

Since it is known that the angles of the infeed constants are close to zero, their values can be calculated using magnitudes only:

$$K_1 = \frac{2232.8 + 1333.8 + 0 + 449.9}{2112.6} = 1.90$$

Therefore, the effective cover along the Juanchito-Yumbo line is

$$X_2 = \frac{8.77 - 7.21}{6.79(1 + 1.90)} = 0.079 = 7.9 \text{ per cent}$$

As expected, the reach of zone 2 is less than 50 per cent of the Juanchito-Yumbo line, since the shortest line is the Juanchito 115 to Juanchito 220 circuit. The reaches of the relays are given in Figure 9.36.

Example 9.2

For the power system in Figure 9.37, calculate the zone 2 setting for the relay at San Antonio with the infeed constant based on the results for a fault in the Chipichape substation, taking into account the criteria stated earlier in this chapter.

Determine the actual impedances that the relay sees for a fault on the Chipichape-Yumbo line and on the busbar at Yumbo and, using these, determine whether zone 2 of the relay operates for these faults.

In calculating the zone 2 reach, check the amount of line covered if the contribution from the Yumbo-Chipichape line is neglected. Work with primary impedances.

Solution

The second zone of the relay extends to the Chipichape-Yumbo line, which is the shortest adjacent line seen by the relay at San Antonio.

In the process, the initial fault point was taken as being at Chipichape, and then moved in 10 per cent steps along one of the Chipichape-Yumbo circuits. With the aid of a computer it was possible to determine the value of K for each point by making a busbar called TEST the fault point. In this way the value of K for each case was calculated using the following expression:

$$K = \left[\frac{(I_{\text{SA-Ch}}/2) + I_{\text{BA-Ch}} + I_{\text{D-Ch}} + I_{\text{Y-Ch}}}{I_{\text{SA-Ch}}/2} \right]$$

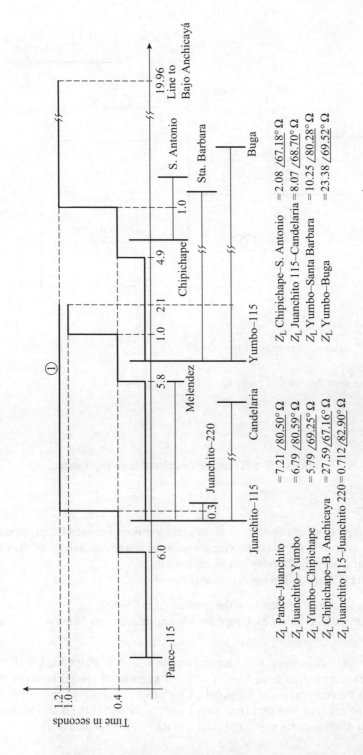

Figure 9.36 Reach of distance protection relays for Example 9.1

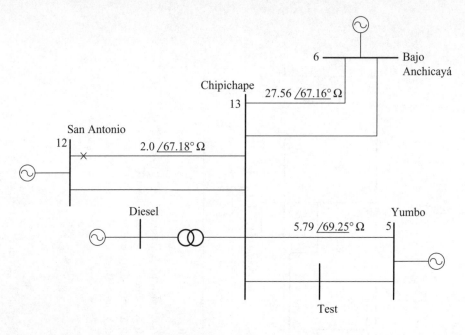

Figure 9.37   Power system for Example 9.2

It can be seen that, in this case, $I_{inf} + I_{relay} = I_{Ch\text{-}TEST}$.
Now

$$\frac{I_{inf} + I_{relay}}{I_{relay}} = K + 1$$

and, therefore, the value of $(1 + K)$ can be calculated from the formula

$$\frac{I_{Ch\text{-}TEST}}{(I_{SA\text{-}Ch})/2}$$

The computer results for a fault at different locations between Chipichape and Yumbo are given in Table 9.1. Knowing the value of K for each fault point, the value of the actual Z seen by the relay can be calculated.

In Figure 9.38 the following curves are shown:

- line impedance as a function of the distance from Chipichape;
- actual impedance from Chipichape seen by the relay at San Antonio;
- value of $(1 + K)$.

The setting values have also been drawn for a fault at Chipichape with, and without, the contribution from Yumbo which is variable. It should be noted that, considering the contribution of the parallel Chipichape-Yumbo line, cover in zone 2 goes up to 68 per cent and then from 94 per cent to the end of the line. This means that overlap could occur at the end of zone 2 of the relay at Chipichape.

Table 9.1 *Values of current and impedance for faults on the Chipichape-Yumbo line*

Location of the fault (%)	$I_{relay} = (I_{SA-Ch})/2$	$I_{Ch-fault}$	I_{Y-Ch} in parallel line	Infeed current	$1+K$	Z_{real} from $Ch = (1+K)Z_L$
0(Ch)	1534.8∠−87°	7748.6∠−83°	3243.5∠−82°	6227.1∠−83°	5.048	0
10	1295.7∠−87°	6117.8∠−85°	2175.9∠−83°	4822.1∠−85°	4.721	2.733
20	1213.0∠−86°	5591.5∠−85°	1858.8∠−83°	4378.5∠−85°	4.609	5.337
30	1135.0∠−86°	5084.6∠−85°	1546.6∠−84°	3949.9∠−85°	4.479	7.780
40	1061.2∠−85°	4594.5∠−85°	1237.8∠−84°	3533.5∠−84°	4.329	10.025
50	990.7∠−85°	4114.6∠−85°	928.1∠−85°	3123.9∠−85°	4.153	12.022
60	922.7∠−84°	3638.2∠−85°	612.7∠−87°	2715.2∠−85°	3.943	13.697
70	8556.2∠−84°	3158.9∠−85°	288.2∠−92°	2302.3∠−85°	3.689	14.951
80	790.6∠−83°	2669.0∠−84°	70.8∠+132°	1878.4∠−84°	3.375	15.633
90	725.5∠−82°	2164.6∠−84°	422.0∠+101°	1439.0∠−84°	2.983	15.544
100(Y)	576.8∠−80°	1115.8∠−81°	1115.8∠+99°	539.8∠−83°	1.934	11.197

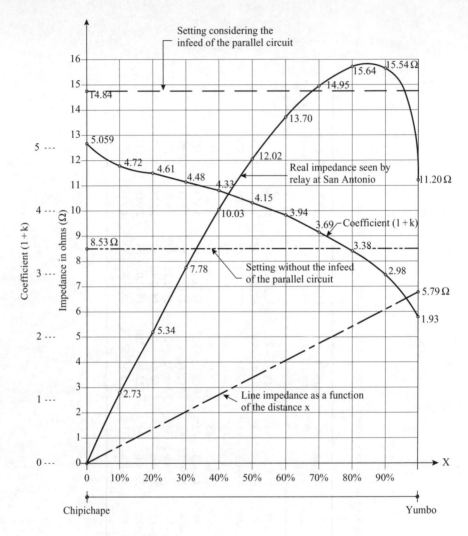

Figure 9.38   Impedance curves for Example 9.2

9.12   Intertripping schemes

Used on high voltage lines for many years, distance protection has proved to be a very reliable form of protection, notwithstanding the limitation that its first zone of coverage does not protect the whole of the circuit, compared to unit protection schemes such as differential protection. With rapid, reliable and economic communication links now available, this limitation can be overcome by providing links between the relays at each end of the line and adding intertripping to distance protection schemes. In addition, distance protection has the advantage of being able to act as back up to protection located at substations further along the line. Although there is a wide variety of intertripping schemes, only the more common ones will be considered here. As with any protection system, the selection of a particular scheme depends on the criteria adopted by individual utilities, the communication links available, and the importance of the circuits being protected. Ultimately it is the fact that these schemes can rapidly clear faults at the far end of the line, outside zone 1, which favours the use of intertripping.

9.12.1   Under reach with direct tripping

In this case, the settings of the distance relay that protects a line follow the criteria referred to in Section 9.3. When the zone 1 units of the relays operate, they initiate a signal that is sent along the communications link to trigger an immediate tripping at the other end of the line. The scheme is simple and has the advantage of being extremely fast; however, it has the disadvantage that it may set off undesirable circuit breaker tripping if there is any mal-operation of the communication equipment. Figure 9.39 illustrates the operation of the scheme with the distance relay located at one end of the line.

9.12.2   Permissive under reach intertripping

This scheme is similar to the one described in the previous subsection, but differs in that the zone 2 unit at the receiving end has to detect the fault as well before the trip signal is initiated. The advantage of this scheme is that spurious trip signals are blocked. Thus, proper account is taken of what has happened to the zone 2 unit of the relay that received the signal from the relay located at the other end. In some cases it is necessary to include a time delay to the tripping command from the remote end, particularly when there are double circuit lines fed by one source located at one end of the lines. Figure 9.40 shows the operation of the scheme with the distance relay located at one end of the line only.

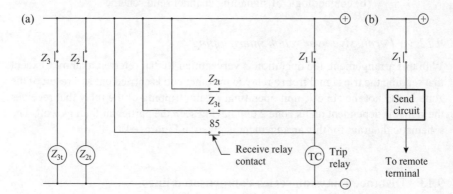

Figure 9.39   Under reach with direct trip: (a) contact logic of trip circuit; (b) contact logic of signalling channel send scheme

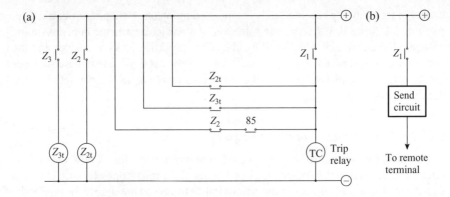

Figure 9.40   Permissive under reach intertripping: (a) contact logic of trip circuit;
(b) contact logic of signalling channel send scheme

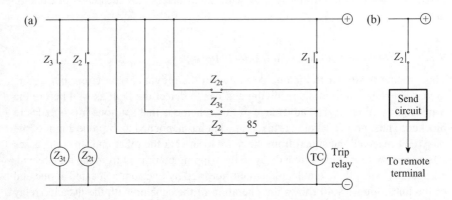

Figure 9.41   Permissive over reach intertripping: (a) contact logic of trip circuit;
(b) contact logic of signalling channel send scheme

9.12.3   *Permissive over reach intertripping*

With this arrangement, the operation is very similar to that referred to above, except
that sending the trip signal from a relay to the other end is carried out as a result of the
zone 2, and not the zone 1, unit operating. Again, tripping of the relay that receives
the signal is dependent on its zone 2 unit having seen the particular fault as well. The
schematic diagram for this arrangement is given in Figure 9.41.

9.13   Distance relays on series-compensated lines

The compensation of lines using series capacitors has proved to be an effective method
of increasing the efficiency of the power transfer along the circuit. The main reasons

for using line compensation are:

- improvement in reactive power balance;
- reduced system losses;
- improved voltage regulation;
- improved system transient stability performance;
- increased power transfer capability.

As regards the last point, the active power transfer from one system (1) to another (2) is given by the expression $P = \{V_1 V_2 \sin(\phi_1 - \phi_2)\}/X$. In the case of a line, the introduction of a series capacitor reduces the overall line reactance and therefore increases the amount of active power that can be transferred. The amount of compensation is usually quoted as the percentage of the line inductive reactance that is compensated by the series capacitor. Values for line compensation are normally within the range of 20 to 70 per cent .

The addition of series compensation can have serious effects on the performance of the protection system, especially distance relays, related to voltage and/or current inversion and the change of impedance seen by the relay. Figure 9.42 shows the apparent impedance seen by a relay at A when 50 per cent series compensation is applied at the middle of the line. Faults beyond the series capacitor appear closer and, therefore, the zone 1 setting should be adjusted to have a smaller impedance setting to avoid over reach. Figure 9.43 corresponds to the case where 70 per cent series compensation is applied at the near end of the line at A. In this case, the relay can see the fault in the reverse direction so that the setting must rely on the memory and biasing of the healthy phases to guarantee proper operation of the protection.

9.14   Technical considerations of distance protection in tee circuits

In the application of distance relays in tee circuits, special attention should be paid to the infeed effect due to the terminals that make up the tee. There can be infeeds at two or three of the terminals, which will require special attention in each case.

9.14.1   Tee connection with infeeds at two terminals

The situation is illustrated in Figure 9.44 where it is assumed that there is no generation at busbar C. The infeed current I_B results in the distance relay at the busbar A seeing an apparent impedance that is greater than the true impedance to the fault point. For a fault at F, the relay at A is supplied by the following voltage:

$$V_A = I_A Z_1 + (I_A + I_B) Z_3 \tag{9.45}$$

Thus, the apparent impedance seen by the relay at A is:

$$Z_A = \frac{V_A}{I_A} = Z_1 + \left[1 + \frac{I_B}{I_A} \right] Z_3 \tag{9.46}$$

$$Z_A = Z_1 + (1 + K_A) Z_A$$

where K_A is defined as the system infeed constant.

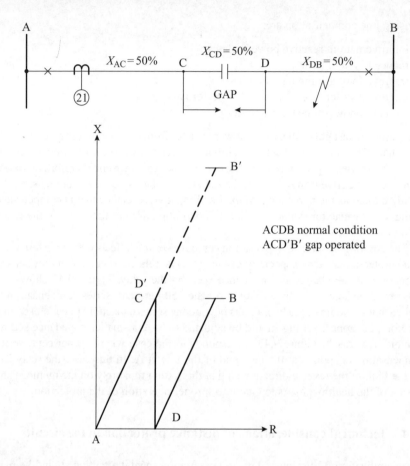

Figure 9.42   Apparent impedance with series compensation at the middle of the line

Since under normal conditions K_A is larger than one, the apparent impedance seen by the relay at A, Z_A, is greater than the actual fault impedance and therefore the relay under reaches the intended cover along the line OC if the infeed is not taken into account. Equally, the relay at B is seeing an apparent impedance given by:

$$Z_B = Z_2 + \left[1 + \frac{I_B}{I_A}\right] Z_3$$

$$Z_B = Z_2 + (1 + K_B) Z_3$$

(9.47)

On the other hand, the relays at A and B should be set in such a way that their zone 1 reaches do not go beyond busbars B and C for the relay at A, and busbars A and C for the relay at B. If not, faults on the transformer in substation C could initiate tripping of the line AB. Zone 1 of the relay at A should be adjusted with the smaller

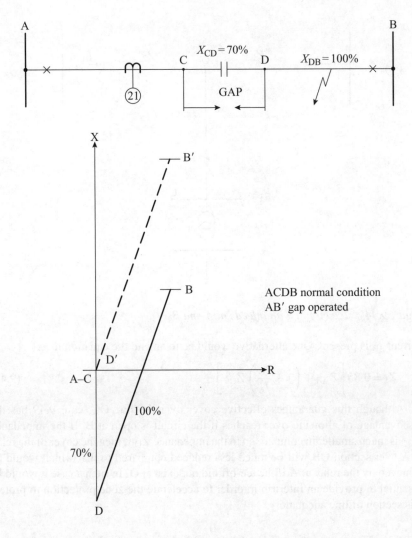

Figure 9.43   Apparent impedance with series compensation at the beginning of the line

of the following values:

$$Z_1 = 0.85Z_{AB}$$
$$Z_1 = 0.85(Z_{AO} + Z_{OC}) \tag{9.48}$$

This guarantees maximum cover over sections OB and OC, without the possibility of over reach from the relay at substation A when the infeed current I_B disappears. However, this will initiate an under reach of the relay under normal conditions when

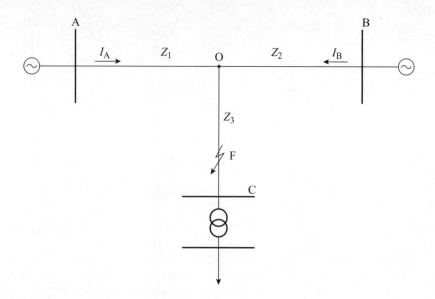

Figure 9.44   Tee circuit with infeeds at A and B

current I_B is present. One alternative would be to amend the equation to

$$Z_1 = 0.85\left[Z_{AB} + \left(1 + \frac{I_B}{I_A}\right)Z_{OC}\right] \tag{9.49}$$

Although this guarantees effective cover over the line OC, eqn. 9.49 has the disadvantage of allowing over reaches if the circuit is open at B. If the impedance Z_{OC} is much smaller in comparison to the impedance Z_{OB}, then the cover of the relay at A over section OB will be much less reduced compared to that which would be achieved by the relay at A if the tee-off did not exist at O. In such a case it would be essential to provide an intertrip in order to accelerate the zone protection to protect this section of line adequately.

9.14.2   Tee connection with infeeds at all three terminals

If a supply infeed exists at the end of the tee-off, as illustrated in Figure 9.45, the infeed effect is still obtained for faults on the line AB resulting in under reach of the relays at A and B. In such a situation, the relay at A would see the following apparent impedance for a fault F, as indicated in Figure 9.45.

$$Z_A = \frac{V_A}{I_A} = Z_1 + \left[1 + \frac{I_C}{I_A}\right]Z_C \tag{9.50}$$

The value of Z_A is greater than the value of the actual fault impedance $(Z_1 + Z_2)$, resulting in under reach in relay A. The settings at A should therefore be calculated on the basis of the actual system impedances, without considering the infeed effect, in order to avoid over reaches when one or more of the terminals of the tee are open.

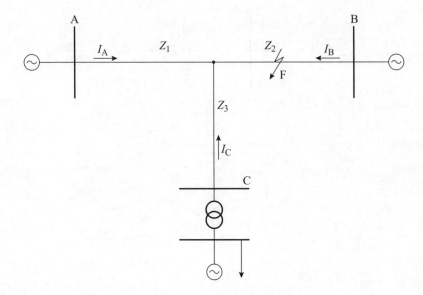

Figure 9.45   Tee circuit with infeeds at A, B and C

With this criterion, optimum co-ordination will be obtained but the reach of the relays is reduced by the effect of the infeed at the terminals.

9.15   Use of distance relays for the detection of the loss of excitation in generators

An excitation fault on a generator produces loss of synchronism with a consequent reduction in power generated and overheating in the windings. The quantity that varies most when a generator loses synchronism is the impedance measured at the terminals of the stator. Under loss of excitation, the voltage at the terminals will begin to fall and the current will increase, resulting in a reduction in this impedance and a change in power factor. The generator and associated power system can be represented as shown in Figure 9.46. The voltage vector diagram is given in Figure 9.47, and the impedance vector diagram in Figure 9.48, for a relay for detecting loss of synchronism located at point A. The impedance seen by the relay, when there are variations in the magnitudes of E_G, E_S, I and δ, are circles with their centres along the straight line CD.

When a generator that is operating synchronously loses excitation, the ratio E_G/E_S decreases and the angle δ increases. This condition, when drawn in the impedance plane, represents a movement of the load point (or of the impedance seen by the relay) in the direction shown in Figure 9.49. A relay with mho characteristics that has only two settings – the offset and the diameter – is used for detecting this condition as shown in Figure 9.50. The offset is designed to prevent operations during system oscillations when excitation has not been lost and protection against asynchronous operation is required. The diameter setting should enable power to be supplied to

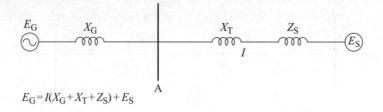

$$E_G = I(X_G + X_T + Z_S) + E_S$$

Figure 9.46   Equivalent system to analyse loss of field

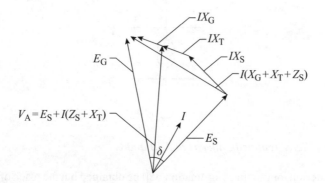

Figure 9.47   Voltage vector diagram for system of Figure 9.46

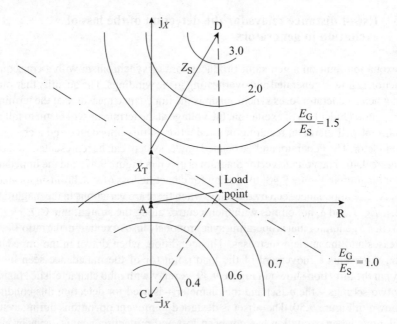

Figure 9.48   Impedance vector diagram for Figure 9.46

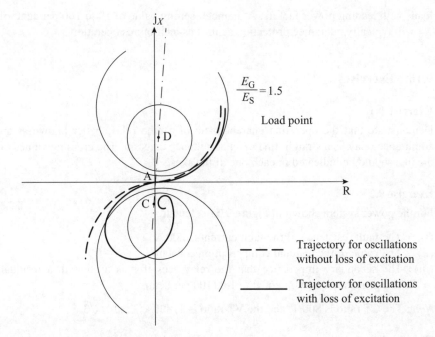

$$\frac{E_G}{E_S} = 1.5$$

Load point

--- Trajectory for oscillations
without loss of excitation

— Trajectory for oscillations
with loss of excitation

Figure 9.49   Load point movement

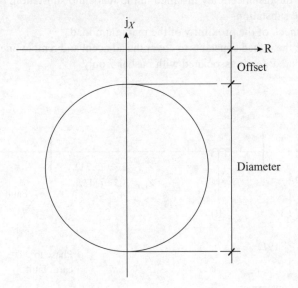

Figure 9.50   Mho relay offset and diameter settings

loads with leading power factors. A diameter setting value of 50 to 100 per cent of X_d will typically guarantee protection against asynchronous operation.

9.16   Exercises

Exercise 9.1

Demonstrate that the operating characteristic of a mho relay, better known as an admittance relay, is a straight line in an admittance diagram. The crossing values of the line should be indicated in each one of the axes.

Exercise 9.2

For the power system shown in Figure 9.51 calculate:

(i)   The fault resistance, if the fault current is 200 A.
(ii)   The value of the residual compensation constant.
(iii) The secondary impedance that the relay sees if it is used with a residual compensation constant equal to 1.0 (100 per cent).

Note: The CT ratio is 800/1, and the VT ratio is $11800/\sqrt{3} : 110/\sqrt{3}$.

Exercise 9.3

For the power system shown in Figure 9.52 determine the reach in secondary ohms for zone 3 of the distance relay installed in the Juanchito substation, on the line that goes to Pance substation.

Make a check of the proximity of the maximum load.

Calculate the infeed constants to cover the adjacent and remote lines, considering the intermediate infeeds associated with busbar 7 only.

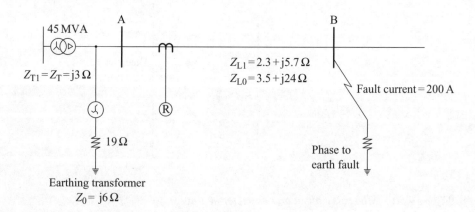

Figure 9.51   Power flow for Exercise 9.2

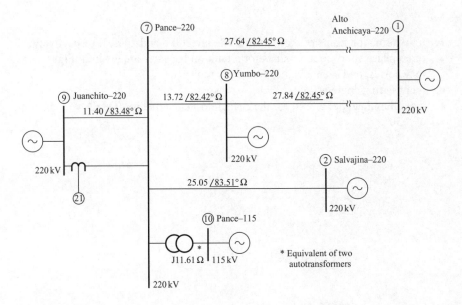

Figure 9.52   *Power system for Exercise 9.3*

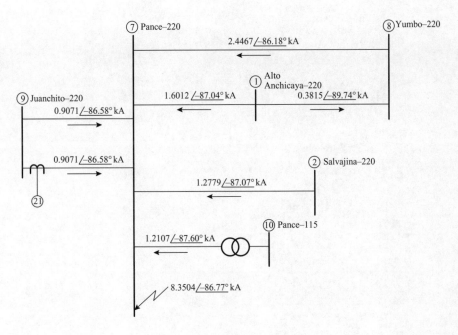

Figure 9.53   *Fault current contribution for Exercise 9.3*

Note:

- transformation ratios of the CTs and VTs are 800/5 and 2000/1, respectively;
- three-phase short-circuit values for a fault on busbar 7 are given in Figure 9.53; they correspond to line values (not per circuit);
- relay setting angle is 75°;
- maximum load per circuit for the Juanchito-Pance line is 40 MVA with an angle of ±30°.

Chapter 10
Protection of industrial systems

With the increase in size of industrial plant electrical systems, and the high short-circuit levels encountered on electricity power systems, it is essential that the electrical protection arrangements in any industrial installation are correctly designed and have the appropriate settings applied to ensure the correct functioning of the plant and continuity of supply within the installation. The importance of maintaining continuity of supply to industrial installations cannot be over emphasised and, in this respect, the interconnectors to the public supply system play a vital role. It is crucial that correct co-ordination is maintained between the protection on the main industrial supply infeeds and the power system supply feeders.

10.1 Protection devices

In addition to the overcurrent relay, which has been covered in Chapter 5, moulded-case circuit breakers (MCCBs) and thermal relay-contactor and fuse devices are frequently used to protect elements of the industrial electrical system and these will be covered in more detail in this chapter.

10.1.1 Overcurrent relays

This type of relay is usually equipped with an instantaneous and/or time-delay unit. This latter unit can be inverse, very inverse or extremely inverse, and when it is necessary to prevent the relay from operating in one direction it should be provided with a directional element. This is required for ring systems or networks with several infeed sources, the latter being very common in industrial systems.

10.1.2 Direct acting devices in power and moulded-case circuit breakers

As their name indicates, these are devices that act directly on power breakers and therefore do not require AC or DC coils for tripping. They are especially common for operating breakers up to 600 V, but are sometimes used on breakers of a higher voltage and, in these cases, are generally fed by current transformers.

These devices can be operated by:

1.   An armature attracted by the electromagnetic force that is produced by the fault current which flows through a trip coil.
2.   A bimetallic strip that is actuated by the heat produced by the fault current.

The characteristic curves of these mechanisms are generally the result of combining the curves of instantaneous relays with long or short time-delay relays. The starting current of the long delay units can normally be set to 80, 100, 120, 140 and 160 per cent of the nominal value. The calibrations of short time-delay relays are typically 5, 7.5 and 10 times the nominal value.

It is common to use breakers to protect circuits of low current capacity with a combination of a single time-delay element plus an instantaneous unit normally having a bimetallic or magnetic element. In this case, the curves are normally set in the factory based on the nominal values.

10.1.3   Combined thermal relay contactor and fuse

The combined thermal relay contactor and fuse is used extensively for protection, mainly in low power systems. In this case the fuse provides protection against short-circuits and the thermal relay gives protection against overloads. Given that the thermal relay acts directly on the contactor, special care should be taken to prevent a relay operation for values of fault current that exceed the capacity of the contactor. If this should be the case, a more rapidly acting fuse should be selected in order to guarantee that it will operate for any current greater than the breaking capacity of the contactor.

10.2   Criteria for setting overcurrent protection devices associated with motors

The criteria normally used for the selection of the nominal values and the range of settings of low voltage overcurrent devices such as thermal relays and moulded-case breakers that are used frequently in industrial plants are similar to those for overcurrent protection included in Chapter 5. An important consideration for these devices is that, as for overcurrent relays, the selected settings can vary depending upon the criteria adopted by the particular utility or plant operators, providing that the resultant settings guarantee appropriate protection to the machines and the elements of the system under analysis. Thus, the settings should be higher than the motor-locked rotor current and below the motor thermal limit.

10.2.1   Thermal relays

A thermal relay basically has three parameters that can be adjusted: the rating of the coil, the range of taps in the thermal element, and the range of the instantaneous element.

Coil rating

In order to determine the coil rating, manufacturers provide a range of maximum and minimum current values for which the thermal relay has been designed. The rating of the coil is somewhat above the maximum value of the motor secondary rated current. The range of a thermal relay for a motor should overlap the motor manufacturer's value of the motor nominal secondary current, in amperes. Using a thermal relay that has a maximum current rating very close to the motor nominal secondary current (in amperes) is not recommended; in this case it is better to use the next highest range available.

Example 10.1

Consider a motor with the following characteristics:

  power: 100 HP, p.f. $= 0.8$
  voltage: 440 V
  efficiency: 100 per cent

The thermal protection consists of three single-phase relays, fed from a set of current transformers with 200/5 ratios. The setting range of the relays is given in Table 10.1, and the operating characteristics are shown in Figure 10.1.

$I_{nom} = 122.36$ A, and $I_{nom(sec)} = 122.36 \times 5/200 = 3.06$ A. In accordance with the data in Table 10.1, a thermal coil with a rating of 3.87 A, which has a current range of 3.10 to 3.39 A, can be selected. A relay with a range of 2.82–3.09 A should not be used since the maximum value is very close to the $I_{nom(sec)}$ of the motor.

Range of taps in the thermal element

The tripping current of the thermal element of the relay is normally specified with a range of 90/95/100/105/110 per cent of the current rating of the coil.

Range of instantaneous elements

The instantaneous elements are specified at ten times the nominal current of the motor. Typically this range is 6–150 A. For this example:

$I_{inst} = 10 \times I_{nom} = 10 \times 122.36 \times 5/200 = 30.59$ A (secondary)

This confirms that the 6–150 A range is appropriate.

10.2.2 Low voltage breakers

The low voltage breakers used to protect motors usually have two elements: a time-delay unit for long-time overloads and an instantaneous element for short-circuits. The short-time element is optional and recommended only for the more powerful motors, or when the possibility exists of losing co-ordination with other breakers located nearer to the source.

The values that should be specified for a breaker are as follows: the nominal current, and the setting ranges for the time-delay unit, the instantaneous unit, and also the short-time unit if this is fitted.

Table 10.1   *Settings for type TMC21B and TMC23B relays*

Adjusted motor full-load current, A		Thermal coil rating, A continuous	Target unit, A 0.2/2	Instantaneous unit, A 6-150	Model numbers Hand reset thermal unit		Case size S-1	Approx. wt. lb(kg)	
min	max				50/60 Hz* standard time	60 Hz† short time		net 12(5.4)	ship 16(7.3)
1.32	1.45	1.64			12TMC21B1A	12TMC23B1A			
1.46	1.59	1.82			B2A	B2A			
1.60	1.75	2.00			B3A	B3A			
1.76	1.93	2.20			B4A	B4A			
1.94	2.11	2.42			B5A	B5A			
2.12	2.33	2.65			B6A	B6A			
2.34	2.55	2.92			B7A	B7A			
2.56	2.81	3.20			B8A	B8A			
2.82	3.09	3.52			B9A	B9A			
3.10	3.39	3.87			B10A	B10A			
3.40	3.74	4.25			B11A	B11A			
3.75	4.11	4.88			B12A	B12A			
4.12	4.47	5.15			B13A	B13A			
4.48	4.97	5.15			B14A	B14A			

* This relay has a time/current curve suitable for use with most general purpose AC motors, 50/60 Hz.
† This relay has a fast tripping time suitable for applications where the motor heats rapidly under stalled conditions.

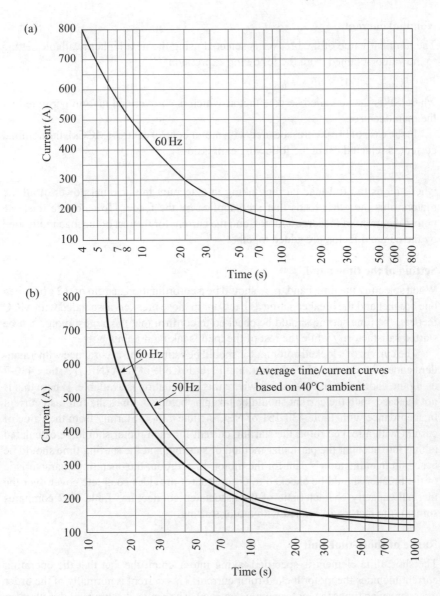

Figure 10.1 *Range of settings and operating characteristics of GE thermal relays TMC.* (a) Typical time/current characteristic curve for type TMC 23B relays. (b) Average time/current curves based on 40 degrees ambient for relays type TMC21B, 24B and 24D. Source: *Protection of electricity distribution systems.* Reproduced by permission of General Electric Company

Nominal current

The nominal value of the breaker is selected using the next higher available setting to the value obtained from the following expression:

$$I_{breaker} = 1.05 \times SF \times I_{nom.motor}$$

where SF is the so-called service factor, which is an overload margin permitted by the manufacturer.

In the case of breakers associated with motor control centres (MCCs), the nominal current is selected using the following expression:

$$I_{nom} = 1.2 \times I_{FL}$$

where I_{FL} is the full-load current taking into account the nominal power of all the motors plus the other loads that are supplied by the feeder. To calculate this, the nominal current of the largest motor, and the current for the rest of the load multiplied by the demand factor, are added together.

Setting of the time band

When selecting the time band there should be a co-ordination margin of 0.2 s between one breaker and the breaker acting as back up. In those breakers associated with MCC feeders, the time setting should be checked to confirm that the largest motor can be started satisfactorily while the rest of the load is taking nominal power.

The current of a locked-rotor motor should be estimated in accordance with a standard code of practice such as the National Electric Code of USA (NEC) Article 430-7 using the code letter in those cases where this information is available. Where this is not the case, then the current should be taken as being six times the full load current in accordance with Table 430-151 of NEC Article 430-7. Starting from the value of current with a locked rotor, the starting current for each motor should be calculated taking into account the particular method of starting, and the starting time should be based on manufacturer's data for the motor. In addition, the operating characteristic of the breaker should be checked to ensure that it provides complete cover over the thermal-capacity characteristic of the associated conductors. Table 10.2 compares some of the factors associated with motor starting.

Range of short-time unit

The short-time element is specified taking into account the fact that the operating current includes the motor locked-rotor current. This current is normally of the order of six or seven times the motor nominal current. The range of settings of this element is usually expressed as a multiple of the starting current selected for the long time-delay unit.

Setting of instantaneous element

The instantaneous element provides protection against short-circuits, cutting down the tripping time of a breaker when there are severe faults on the associated circuit. The setting of the instantaneous element is calculated using the expression $I_{inst} = 10 \times I_{FL}$, where I_{FL} is the full load current of the associated feeder.

Table 10.2   Comparison of motor starting methods (from Industrial Power Systems Handbook, *by D. Beeman, 1985; reproduced by permission of McGraw-Hill Publishing Company)*

Type of starter*	Motor voltage	Starting torque	Line current
	Line voltage	Full voltage starting torque	Full voltage starting current
Full-voltage starter	1.0	1.0	1.0
Autotransformer			
80% tap	0.80	0.64	0.68
65% tap	0.65	0.42	0.46
50%	0.50	0.25	0.30
Resistor starter, single step (adjusted for motor voltage to be 80% of line voltage)	0.80	0.64	0.80
Reactor			
50% tap	0.50	0.25	0.50
45% tap	0.45	0.20	0.45
37.5% tap	0.375	0.14	0.375
Part-winding starter (low speed motors only)			
75% tap	1.0	0.75	0.75
50% tap	1.0	0.50	0.50

line voltage = motor rated voltage
* The settings given are the more common for each type.

Example 10.2

Determine the settings for the thermal relay and the 200 and 600 A breakers, which protect the system indicated in Figure 10.2, using the information given.

Induction motor

500 HP, 2400 V, power factor = 0.8
Service factor: 1.0
Code letter: G; thermal limit with locked rotor: 5.5 s
Direct start, duration: 1.0 s

Thermal relay
(See Figure 10.1.)

Coil: 3.87 A
Taps: 90–110 per cent of coil rating
CT ratio: 150/5

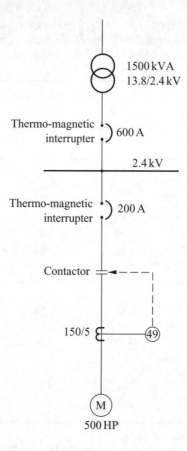

Figure 10.2   System for Example 10.2

Breakers

The setting values and characteristic curve are shown in Table 10.3 and Figure 10.3 respectively.

Solution

Nominal motor current

$$I_N = \frac{500 \times 0.746\,\text{kW}}{\sqrt{3} \times 0.8 \times 2.4\,\text{kV}} = 112.16\,\text{A}$$

Locked rotor current

From NEC Table 430-7(b), a motor with code G is assessed at 6.29 kVA/HP, and so:

$$I_{LR} = \frac{500 \times 6.29\,\text{kVA}}{\sqrt{3} \times 2.4\,\text{kV}} = 756.57\,\text{A}$$

Table 10.3 Range of settings of Siemens Allis breaker

Breaker type and frame size	Tripping XFMR rating (primary A)	Long-time element calibrated pick-up settings (A)							Maximum continuous rating (A)	Ground-element calibrated pick-up settings (A)			
		A	B	C	D	E	F	G		15%	25%	50%	100%
LA-600 A	80	40	50	60	70	80	90	100	100	×	×	40	80
and	200	100	125	150	175	200	225	250	250	30	50	100	200
LA-800 A	400	200	250	300	350	400	450	500	500	60	100	200	400
	600	300	375	400	525	600	675	750	600*	90	150	300	600
LA-800 A only	800	400	500	600	700	800	900	1000	800	120	200	400	800
LA-1600 A	200	100	125	150	175	200	225	250	250	–	50	100	200
	400	200	250	300	350	400	450	500	500	60	100	200	400
	800	400	500	600	700	800	900	1000	1000	120	200	400	800
	1600	800	1000	1200	1400	1600	1800	2000	1600	240	400	800	1600
LA-3200 A	2000	1000	1250	1500	1750	2000	250	2500	2500	300	500	1000	2000
	3200	1600	2000	2400	2800	3200	3600	4000	3200#	480	800	1600	3200
LA-4000 A	4000	2000	2500	3000	3500	4000	4500	5000	4000	600	1000	2000	4000

* 750 on LA-800 A
# 3000 for LAF-3000 A
× may not trip

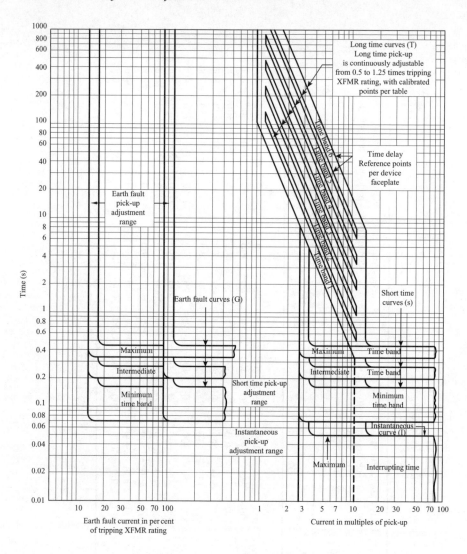

Figure 10.3   Characteristic operating curves for Siemens-Allis Interrupter

Setting of thermal relay

$$I_{start} = 1.05 \times 1.0 \times 112.16\,A = 117.8\,A$$

$$\text{Setting} = 117.8 \times \frac{5}{150} \times \frac{1}{3.87} \times 100\% = 101.46\%$$

With a setting of 100 per cent,

$$I_{start} = 1.0 \times 3.87\,A \times (150/5) = 116.10\,A\ (3.5\%\ \text{overload})$$

Setting of 200 A breaker

(i) Long-time element
Range: 100/125/150/175/200/250 A

$$I_{start} = 1.05 \times 1.0 \times 112.16\,A = 117.8\,A$$

Set at 125 A (11.45 per cent, overload)
Selection of time band:
The long-time element should permit the motor to start:
Start point: $I = 756.57\,A$ locked rotor current, and $t = 1.0\,s$
Operation of breaker at the start:
$I/I_{start} = 756.7\,A/125\,A = 6.05$ times.
Require 1.4 s tripping time. From Figure 10.3, the intersection of 1.4 s and 6.05 times
lies above the lower curve of Band 1. Therefore, Band 2 is chosen in this case to
guarantee the required discrimination margin of 0.4 s.

(ii) Short-time element
Range: 3/5/8/12 times long-time pick-up current

$$I_{start1} = 6 \times (112.16\,A) = 672.96\,A$$

Given that this value is less than the motor starting current, it is necessary to increase
the setting value. Try eight times, then

$$I_{start2} = 8 \times (112.16\,A) = 897.28\,A \text{ and setting} = 897.28\,A/125\,A = 7.18$$

Setting selected: $8 \times I_{pickup} = 8 \times 125\,A = 1000\,A$
Time band: in this case the intermediate band was chosen to provide the necessary
discrimination margin with the instantaneous unit (see Figure 10.3).

(iii) Instantaneous element
Range: 3/5/8/12 times long-time pick-up current
$I_{start} = 12 \times (112.16\,A) = 1345.92\,A$
Setting $= 1345.92\,A/125\,A = 10.77$
Setting selected: $11 \times I_{pickup}$ (1375 A)

Setting of 600 A breaker

(i) Long-time element
Range: 300/375/400/525/600/750 A
Nominal current of transformer, $I_n = \dfrac{1500\,kVA}{\sqrt{3} \times 2.4\,kV} = 360.84\,A$
With a setting of 400 A, the overload $= 400/360.84 = 1.108$, i.e. 10.8 per cent, which
is acceptable.

$$I_{start} = 1.10 \times (360.84\,A) = 396.92\,A$$

Selection of time band:
Operation of motor breaker within the limit of the long-time element:
$I/I_{start} = 1000\,A/125\,A = 8.0$ times. At eight times, and with Band 2, $\Rightarrow t = 1.5\,s$

Operation of transformer breaker:

$I/I_{start} = 1000\,A/400\,A = 2.5$ times. At 2.5 times, and with $t = 1.7\,s$, Band 1 is chosen since this is the lowest available. Notice that a margin of 0.2 s has been applied here since the co-ordination is between two low voltage breakers where the curves include the opening time.

(ii) Short-time element

$$I_{start} = 6 \times (360.84\,A) = 2165.04\,A$$

$$\text{Setting} = 2165.04\,A/400\,A = 5.41 \Rightarrow 6 \text{ times}$$

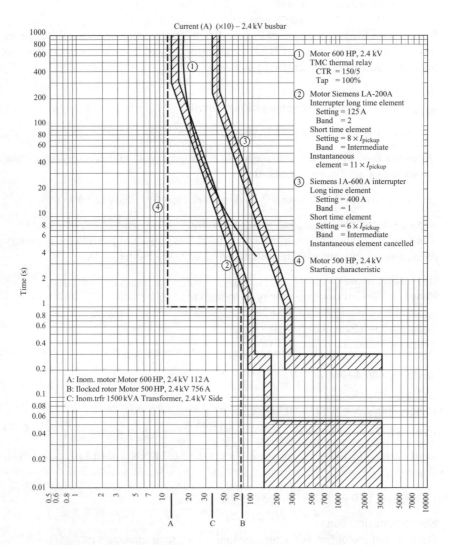

Figure 10.4   Co-ordination curves for Example 10.2

Since there is no overlap with the short-time element of the motor, the intermediate band is also selected.

(iii) Instantaneous element

The instantaneous element is cancelled in order to maintain co-ordination. The co-ordination curves are shown in Figure 10.4.

It should be emphasised that, although the values selected for the protective devices in this example ensure proper co-ordination, they are not unique. Other settings could be chosen provided that the curves so obtained guarantee adequate reliability and good selectivity.

Chapter 11

Industrial plant load shedding

All electrical power systems that contain generation are liable to be subjected to a variety of abnormal operating conditions such as network faults, loss of some or all generation, the tripping of circuits within the system and other disturbances that can result in the reduction in the generation capacity available to the system. In these situations a balance between the existing load and the remaining generation should be re-established if possible before the reduction in frequency produced by the overload affects the turbines and the generation auxiliary equipment, which could eventually result in the total collapse of the system.

To assist in restoring the equilibrium, frequency relays are employed. These disconnect the less important loads in stages when the frequency drops to a level that indicates that a loss of generation or an overload has occurred. This type of scheme is very useful in industrial plants where in-house generation is synchronised to the public grid system. The concepts, criteria and conditions applicable to the design of an automatic load shedding system for industrial plants, based on frequency relays, are set out in this chapter.

11.1   Power system operation after loss of generation

When a total or partial loss of generation occurs within the system, the first indicators are a drop in voltage and in frequency. However, given that voltage drops can also be caused by system faults, it is generally recognised that a drop in frequency is a more reliable indication of loss of generation. A sudden loss of generation in the system will result in a reduction in the frequency at a rate of change that depends on the size of the resultant overload and the inertia constant of the system.

The relationship that defines the variation of frequency with time, following a sudden variation in load and/or generation, can be obtained, starting from the equation for the oscillation of a simple generator:

$$\frac{GH}{\pi f_0} \times \frac{d^2\delta}{dt^2} = P_{\mathrm{A}} \tag{11.1}$$

where: G = nominal MVA of machine under consideration; H = inertia constant (MWs/MVA = MJ/MVA); δ = generator torque angle; f_0 = nominal frequency; P_A = net power accelerated or decelerated (MW).

The speed of the machine at any instant (W) can be given by the following expression:

$$W = W_0 + \frac{d\delta}{dt} = 2\pi f \tag{11.2}$$

in which W_0 is the synchronous speed, i.e. the nominal speed at rated frequency.

Differentiating eqn. 11.2 with respect to time:

$$\frac{dW}{dt} = \frac{d^2\delta}{dt^2} = 2\pi \frac{df}{dt} \tag{11.3}$$

Replacing eqn. 11.3 in eqn. 11.1 gives

$$\frac{df}{dt} = \frac{P_A f_0}{2GH} \tag{11.4}$$

Eqn. 11.4 defines the rate of the variation of the frequency in Hz/s, and can be used for an individual machine or for an equivalent that represents the total generation in a system. In such a case the inertia constant can be calculated from

$$H = \frac{H_1 MVA_1 + H_2 MVA_2 + \cdots + H_n MVA_n}{MVA_1 + MVA_2 + \cdots + MVA_n} \tag{11.5}$$

where the subscripts $1, 2 \ldots n$ refer to the individual generator units. It should be emphasised that the constant H in eqn. 11.5 is expressed to an MVA base equal to the total generation capacity of the system.

The accelerating power P_A in eqn. 11.4 is responsible for the frequency variation. It can be calculated from

$$P_A = P_M - P_E \tag{11.6}$$

where P_M = the mechanical power entering the generator and P_E = the electrical power leaving the generator.

Under stable conditions $P_A = 0$ and there are no frequency variations. In the case of overloads $P_E > P_M$. Thus, $P_A < 0$ and there will be a drop in the system frequency.

11.2 Design of an automatic load shedding system

In order to design an automatic load shedding system, a model that represents the different generating machines should first be defined, and then the load parameters and the criteria for setting the frequency relays.

11.2.1 Simple machine model

Within the scope of this book, a single machine has been used in the power system model used to illustrate a load shedding system. This is equivalent to assuming that the generator units are electrically connected with negligible oscillations between them, and with a uniform frequency across the whole of the system, ignoring the effect of the regulating equipment. The load is represented as a constant power, which implies that there is no reduction in load as a result of the voltage and frequency drops after a contingency situation. This model provides a pessimistic simulation of the system since the reduction of the load due to the frequency drops and the effect of the speed regulators are neglected. In using this model the inertia constant of the system is calculated using eqn. 11.5.

The rate of change of the frequency is calculated from eqn. 11.4 with the following assumptions:

- the mechanical power entering the generators does not vary and is equal in magnitude to the prefault value;
- the magnitude of the load does not vary with time, voltage or frequency. It is only reduced by disconnecting part of the load as a result of the automatic load shedding system.

This simple machine model, with loads modelled as a constant power, is used to determine the frequency relay settings and to verify the level of minimum frequency attained before a contingency situation is reached, for the following reasons:

- the ease of using an iterative process to design the load shedding system;
- consideration of suitable security margins, since the fact that the load diminishes with loss of voltage is neglected. This implies introducing much more severe rates of frequency variation, thus achieving more rapid settings.

11.2.2   Parameters for implementing a load shedding system

The following aspects need to be defined in order to implement the load shedding system.

Maximum load to be disconnected

Generally, one of the most drastic conditions corresponds to the total loss of interconnection between the public electricity supply network and the internal electrical system within the industrial plant. In this case, the unbalance between generation and load will be equal to the maximum import and should be compensated by the disconnection of a similar amount of load from the plant system.

Starting frequency of the load shedding system

The disconnection system should be set so that it will initiate operation at a value of frequency below the normal working system frequency. Taking into account variations in frequency caused by oscillations inherent in the public system, this value is normally selected at approximately 93 per cent of nominal system frequency. However, if it is thought that there is a possibility of more severe oscillations occurring on the system, then it is recommended that a supervisory control arrangement using

Table 11.1   Typical times for the operation of turbines (full load)

% of rated frequency at full load	Maximum permissible time (minutes)
99.0	continuously
97.3	90
97.0	10
96.0	1

Note: The operating times at low frequency and full load are cumulative.

overcurrent relays, which can detect the outages of circuits connecting the industrial plant to the public system, should be installed to avoid incorrect operations.

Minimum permissible frequency

A steam turbine is designed so that, when operating at nominal mechanical speed and generating at nominal system frequency, excessive vibrations and stresses in its components, e.g. resonance of turbine blades, are avoided. However, when running below normal speed at a reduced system frequency, cumulative damage could be produced by excessive vibration. It is recommended, therefore, that the time limits given in Table 11.1 should not be exceeded. However, during transient operation and with load below nominal, in the majority of cases reduction of frequency down to 93 per cent of rated frequency can be permitted without causing damage either to the turbine or to the turbogenerator auxiliary lubrication and cooling systems.

11.3   Criteria for setting frequency relays

The determination of the frequency relay settings is an iterative process, and is carried out in such a way that the final settings satisfy the requirements of both speed and co-ordination. In this process, the co-ordination between relays that trip successive stages of load should be checked in order to ensure that the least amount of load is shed, depending on the initial overload condition.

11.3.1   Operating times

When selecting the settings, it is necessary to consider the time interval between the system frequency decaying from the relay pick-up value to the point in time when the load is effectively disconnected. The relay pick-up time is included in this time interval, plus the preset time delay of the relay, if this is required, and the breaker opening time.

The following values are typically used for industrial systems: relay pick-up time: 50 ms and breaker opening time: 100 ms.

11.3.2   Determination of the frequency variation

The frequency variation required to calculate the settings is obtained by using a simple machine model for the system, and a constant power model for the load. This assumes that the load connected to the generators is the same before, and after, the contingency, neglecting any form of damping. Given this, the calculated rate of loss of frequency in the system is pessimistic, and the settings determined on this basis thus provide an arrangement that rapidly restores the frequency to its normal value, thereby ensuring a secure system.

11.4   Example of calculating and setting frequency relays in an industrial plant

The procedure for calculating the settings of the frequency relays at a typical industrial plant is given next. The single-line diagram is shown in Figure 11.1, together with the principal data for the study.

11.4.1   Calculation of overload

The initial load conditions are summarised as: total load: 24.0 MW; in-house generation: 8.0 MW; total import: 16.0 MW; GH constant: 35.43.

In the eventuality of the loss of the incoming grid supply, the in-house generators will experience the following overload:

$$\%\text{overload} = \frac{24.0 - 8.0}{8.0} \times 100\% = 200\%$$

11.4.2   Load to be shed

With the loss of the grid supply, 16 MW of capacity is lost, which will have to be borne initially by the in-house generators. A load equal to, or greater than, this amount must be disconnected in order to relieve the overload. Therefore, the load to be shed is $24 - 8 = 16$ MW. It should be noted that there are high priority loads totalling 8.02 MW that cannot be disconnected.

11.4.3   Frequency levels

Disconnection of load is initiated each time the system frequency falls to 59 Hz, which indicates that a loss of generation has taken place. A minimum frequency of 56 Hz is acceptable.

11.4.4   Load shedding stages

Assume that three stages of shedding have been set up that will off load the incoming circuit by 15.98 MW, as detailed in Table 11.2. With this arrangement, the in-house generation can then supply the 8.02 MW of high priority load.

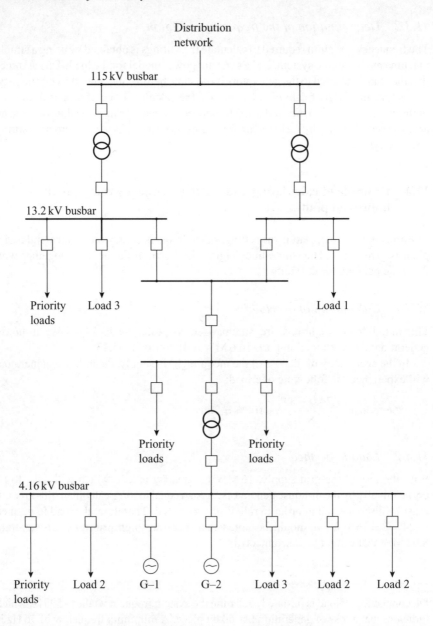

Figure 11.1   System arrangement for example of calculating frequency relay settings

11.4.5   Determination of the frequency relay settings

The settings of the frequency relays shown in Table 11.2 are determined in such a way that each stage is disconnected only when the system frequency falls to a predetermined value. This value is obtained by calculating the reduction in the

Table 11.2   Load shedding stages

Priority	Description	MW
1	load 1	6.03
2	load 2	4.73
3	load 3	5.22

Total load to be shed $= 15.98\,\text{MW}$

system frequency due to an overload equal to the stage considered, as described below.

First-stage setting

The first stage is disconnected when the frequency reaches 59 Hz.

Second-stage setting

An overload equal to the first stage is considered and the subsequent rate of frequency drop is determined from

$$\frac{df}{dt} = \left(\frac{-6.03}{2GH}\right) \times 60 = -5.106\,\text{Hz/s}$$

The frequency as a function of time is given in Figure 11.2 and $f = (60 - 5.106t)$.
    The opening time for the first stage is

$$t_{\text{trip}} = t_{\text{pick-up}} + t_{\text{breaker}} + t_{\text{relay}}$$

$$t_{\text{pick-up}} = (60 - 59)/5.106 = 0.196\,\text{s}$$

$$t_{\text{trip}} = 0.196 + 0.100 + 0.05 = 0.346\,\text{s}$$

The frequency drop up to the operation of the first stage is $f = [60 - 5.106(0.346)] = 58.233\,\text{Hz}$. The second stage is set below this value, i.e. at 58.15 Hz.

Third-stage setting

These are set so that they will not operate for overloads below $(6.03 + 4.73) = 10.76\,\text{MW}$, which are disconnected by stages 1 and 2.

$$\frac{df}{dt} = \left(\frac{-10.76}{2GH}\right) \times 60 = -9.111\,\text{Hz/s}$$

$$f = 60 - (9.111)t$$

The pick-up time for the first stage is $t = (60 - 59)/9.111 = 0.110\,\text{s}$, and the tripping time for this stage is $t_{\text{trip1}} = 0.110 + 0.05 + 0.100 = 0.260\,\text{s}$. Thus, the frequency drop, $f = 60 - 9.111(0.26) = 57.631\,\text{Hz}$.

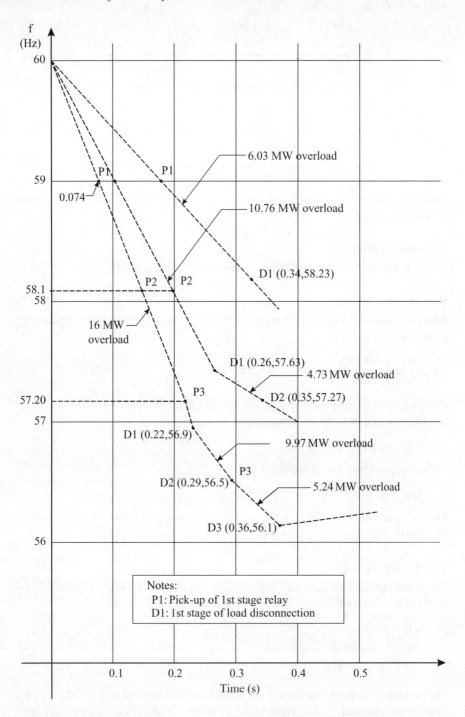

Figure 11.2   Calculation of settings of frequency relays

The value that produces pick-up at the second stage is $t = (60 - 58.15)/9.111 = 0.203$ s, so that $t_{\text{trip2}} = 0.203 + 0.05 + 0.100 = 0.353$ s.

The frequency equation shows that in $t_{\text{trip1}} = 0.260$ s there is a variation of slope as a consequence of the disconnection of the first stage (6.03 MW); see Figure 11.2. After this, the accelerating power is only 4.73 MW.

Therefore

$$\frac{df}{dt} = \frac{-4.73}{2GH} \times 57.631 = 3.847 \, \text{Hz/s}$$

and, from the frequency equation:

$$f - 57.632 = -3.847(t - 0.26), \qquad t > 0.26 \, \text{s}$$

$$f = 58.632 - 3.847t$$

The frequency drop up to the stage 2 tripping is given by $f = 58.632 - 3.847 t_{\text{trip2}}$, from which $f = 57.274$ Hz. Therefore a third stage setting of 57.20 Hz is selected.

The final settings are given in Table 11.3.

The operation of the system in the presence of a total loss of connection with the distribution network is represented by the lower characteristic in Figure 11.2.

11.4.6 Verification of operation

In order to verify the operation of the proposed system, the reduction in frequency during the process of load shedding is studied by two different methods: modelling the load as a constant power, and modelling the load with damping as a result of the voltage drop. For the latter case a transient stability program is used.

Modelling load as a constant power

For this case it is assumed that the magnitude of the load is constant and therefore does not depend on the voltage level. This consideration is pessimistic with regard to the actual situation and therefore provides some margin of security. The frequency analysis is carried out using eqn. 11.4, the settings of the frequency relays given in Table 11.3 and assuming a maximum circuit breaker opening time of 100 ms.

Figure 11.3 shows the behaviour of the frequency from $t = 0$ s, when loss of supply occurs at a total system load of 24 MW. Under such conditions the sequence of events

Table 11.3   Frequency relay settings

Stage	Frequency setting (Hz)	Delay time (s)
1st	59.00	instantaneous
2nd	58.15	instantaneous
3rd	57.20	instantaneous

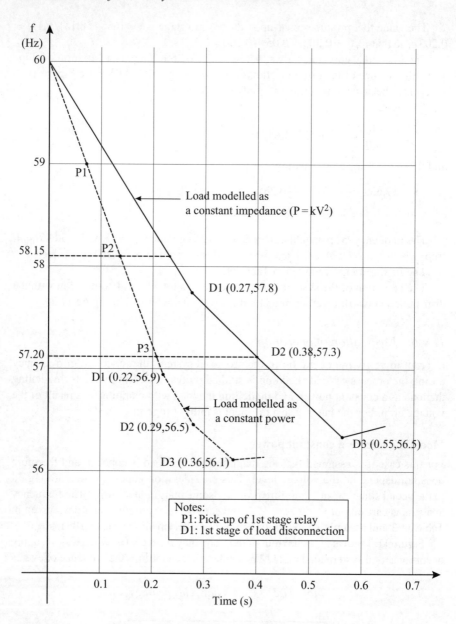

Figure 11.3   Variation of frequency during load shedding

is as given in Table 11.4. The drastic initial overload is eventually eliminated by the load shedding system in 0.357 s. During this time, the frequency will fall to 56.17 Hz, a value which more than meets the standards accepted for industrial systems operating separately. After $t = 0.357$ s, the system frequency then starts to recover.

Table 11.4   Disconnection of the grid supply; sequence of events

Time	Frequency (Hz)	Event	Rate of change of frequency (Hz/s)	Rate of change of load shed (MW)	Overload remaining on generators (MW)
0.000	60.00	Disconnection of grid supply	−13.46	–	16.00
0.074	59.00	Pick-up of 1st stage relay	−13.45	–	16.00
0.137	58.15	Pick-up of 2nd stage relay	−13.45	–	16.00
0.207	57.20	Pick-up of 3rd stage relay	−13.45	–	16.00
0.224	56.90	1st stage of load disconnection	−8.00	6.03	9.97
0.287	56.50	2nd stage of load disconnection	−4.20	4.73	5.24
0.357	56.10	3rd stage of load disconnection	0.00	5.22	0.02

To obtain the values in Table 11.4, it must be borne in mind that, since the initial overload is equal to 16 MW, then

$$\frac{df}{dt} = \frac{-16}{2GH} \times 60 = -13.45\,\text{Hz/s}$$

Therefore, the pick-up time for the first stage is

$$t_p = \frac{60-59}{13.45} = 0.074\,\text{s}$$

The opening time of the breakers associated with the first stage is $t_d = 0.074 + 0.100 + 0.05 = 0.224$ s and the frequency at that moment is $f = 60 - 13.45 t_d = 56.9$ Hz. The values for the other stages can be calculated in a similar way, and are given in Table 11.4.

Modelling load with damping as a result of the voltage drop ($P = kV^2$)

The stability program should be used to set the initial condition of the load defined for this study, as this can simulate the disconnection of the infeed at $t = 0$ s, and subsequently the disconnection of the loads at the time at which the frequency relays are set to operate.

Figure 11.3 also illustrates the frequency characteristic obtained by computer modelling the loads by constant admittances. This method is less drastic than the constant power model since, in this case, the power of the load is damped by the voltage drop. As the overload is reduced, the frequency drop is less than that obtained using the first model. As seen in Figure 11.3, the frequency relay settings, and therefore the tripping of the breakers, is much slower although the minimum value of the frequency eventually reaches 56.55 Hz, which is above the predetermined minimum limit of 56 Hz. From Figure 11.3 and the results of the stability program it can be seen that the frequency after the loss of the third stage gives recovery at a rate of 0.67 Hz/s. This implies that it should take approximately 2.345 s for recovery to the normal level of 60 Hz.

The modelling of the loads by these two methods makes it possible to obtain graphs of the frequency variation as a function of time that correspond to the most adverse and most favourable extremes in the system. In this way, the curves obtained for the two models delineate the operating area of the system under study, before the loss of connection with the grid supply network.

Analysis of voltage with total loss of infeed supply and operation of the load shedding system

The loss of generation in the system not only causes loss of frequency but also a drop in voltage. The automatic load shedding scheme should prevent system voltages falling to such a level as to cause tripping of the contactors on the motors serving the plant.

A check should be made on the voltage levels on the system as follows:

1. Determine the initial voltage at each busbar, using a load flow program.
2. Determine the variations in voltage at each busbar after the loss of connection with the grid supply, and while the load shedding scheme is in operation. For this a transient stability program is used to obtain the voltages at each busbar for each stage of the analysis.
3. Produce the curves of voltage versus time for the busbars feeding the priority loads.

Chapter 12

Protection schemes and substation design diagrams

Previous chapters have detailed the make up and operating characteristics of various types of protection relays. This chapter considers the combination of relays required to protect various items of power system equipment, plus a brief reference to the diagrams that are part of substation design work. A general knowledge of these diagrams is important in understanding the background to relay applications.

12.1 Protection schemes

It is difficult to define precisely the protection schemes that should be adopted for an electricity distribution system, given the large number of valid alternatives for each situation, but some schemes will be presented as a guide for protecting the various elements that make up a power system. However, any protection scheme should strike a balance between the technical and economic aspects so that, for example, sophisticated protection devices are not used for small machines or for less important power system elements.

12.1.1 Generator protection

Generator protection should take into account the importance of the generator and its technical characteristics such as power, voltage and earthing arrangement, plus any economic considerations. A complex protection scheme can ensure that the generator is protected against whatever faults may occur. However, it is unlikely that such a cost could be justified for every generating station, especially those with small units. It is, therefore, necessary to define a protection scheme that is adequate for the size of the machine.

Two generator protection schemes are given below, based on suggestions by manufacturing companies.

Small generators

For small generators, typically up to 5 MVA, it is considered necessary to have:

- protection against internal faults;
- back-up protection for external faults using overcurrent relays with voltage restraint;
- reverse-power protection;
- earth-fault protection, using an overcurrent relay;
- protection against overloads by means of thermal relays.

This scheme is illustrated in Figure 12.1 (for relay identification, see section 3.1.4).

Large generators

For large generators, say over 5 MVA, the protection, which is shown in Figure 12.2, should normally comprise:

- differential protection to cover internal faults;
- earth-fault protection using high impedance relays;
- back-up protection by means of distance or overcurrent protection with voltage restraint;
- reverse-power protection;
- negative-phase sequence protection;
- protection against loss of excitation;
- protection against overload using thermal relays.
- out of step
- inadvertent energisation (50/27)
- stator earth protection (59N and 27N)
- over/under frequency
- over/under voltage

12.1.2 *Motor protection*

The amount of protection, and its type, used for a motor is a compromise between factors such as the importance of the motor, the potential dangers, the type of duty, and the requirements of protection co-ordination against the cost of the protection scheme. The schemes that are illustrated represent common practice and international recommendations for the protection of motors of different powers, and are divided into four categories:

- protection of low power motors (less than 100 HP);
- protection of motors up to 1000 HP;
- protection of motors greater than 1000 HP;
- additional protection for synchronous motors.

In the diagrams the starting equipment of the respective motors has not been represented.

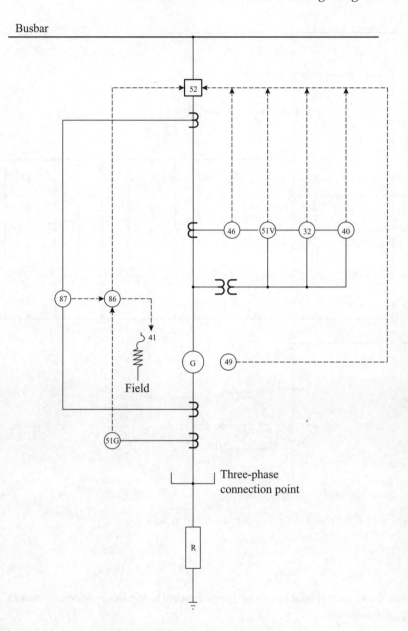

Figure 12.1   Protection schematic for small generators

Protection of low power motors

Low power motors are normally protected by fuses associated with thermal overload relays incorporating bimetallic elements (Figure 12.3) – the fuses protecting against short circuits – or low voltage breakers plus thermal overload relays (Figure 12.4)

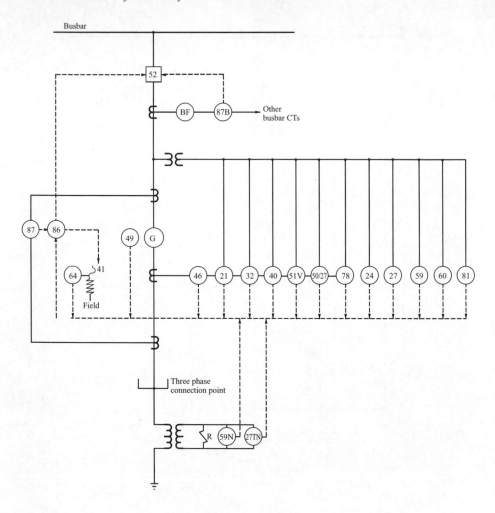

Figure 12.2   Protection schematic for large generators

when the breaker should have a magnetic element to trip instantaneously under short-circuit conditions.

Protection of motors up to 1000 HP

The protection arrangements should include thermal protection against overloads and short-circuits (49/50), protection for locked rotor (51) and earth-fault protection (50G), as indicated in Figure 12.5.

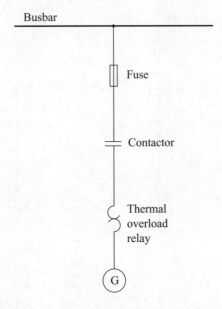

Figure 12.3   Schematic of fuse protection for low power motors

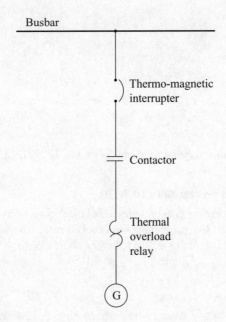

Figure 12.4   Schematic of low voltage breaker protection for low power motor

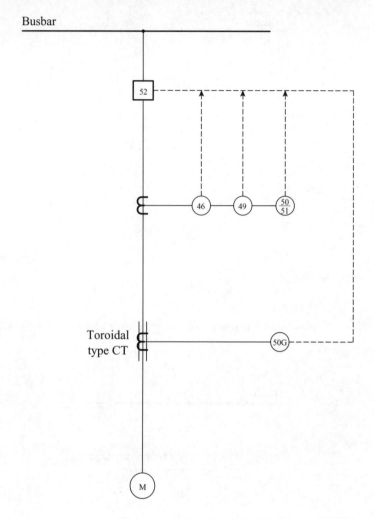

Figure 12.5   Protection schematic for motors up to 1000 HP

Protection of motors greater than 1000 HP

The scheme shown in Figure 12.6 includes unbalance protection (46), thermal protection against overloads (49), protection for a locked-rotor situation (51), differential protection for internal faults (87), back up for short-circuits (50), and earth-fault protection (50G).

Additional protection for synchronous motors over 1000 HP

In addition to the protective devices indicated in Figures 12.4 and 12.5, a large synchronous motor requires protection for the field winding, plus a low power factor relay (55) and undervoltage protection (27), and a high/low frequency relay (81)

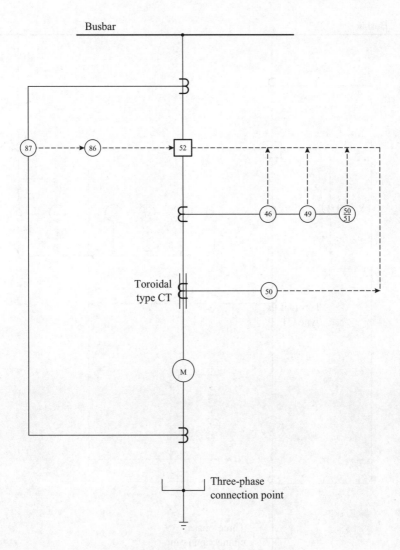

Figure 12.6   Protection schematic for motors over 1000 HP

to prevent the motors running under conditions of low frequency operation. The schematic diagram for the protection of synchronous motors over 1000 HP is given in Figure 12.7.

Protection for the field winding

This would require an earth-fault protection relay (64), and a field relay (40) to deal with loss of excitation current.

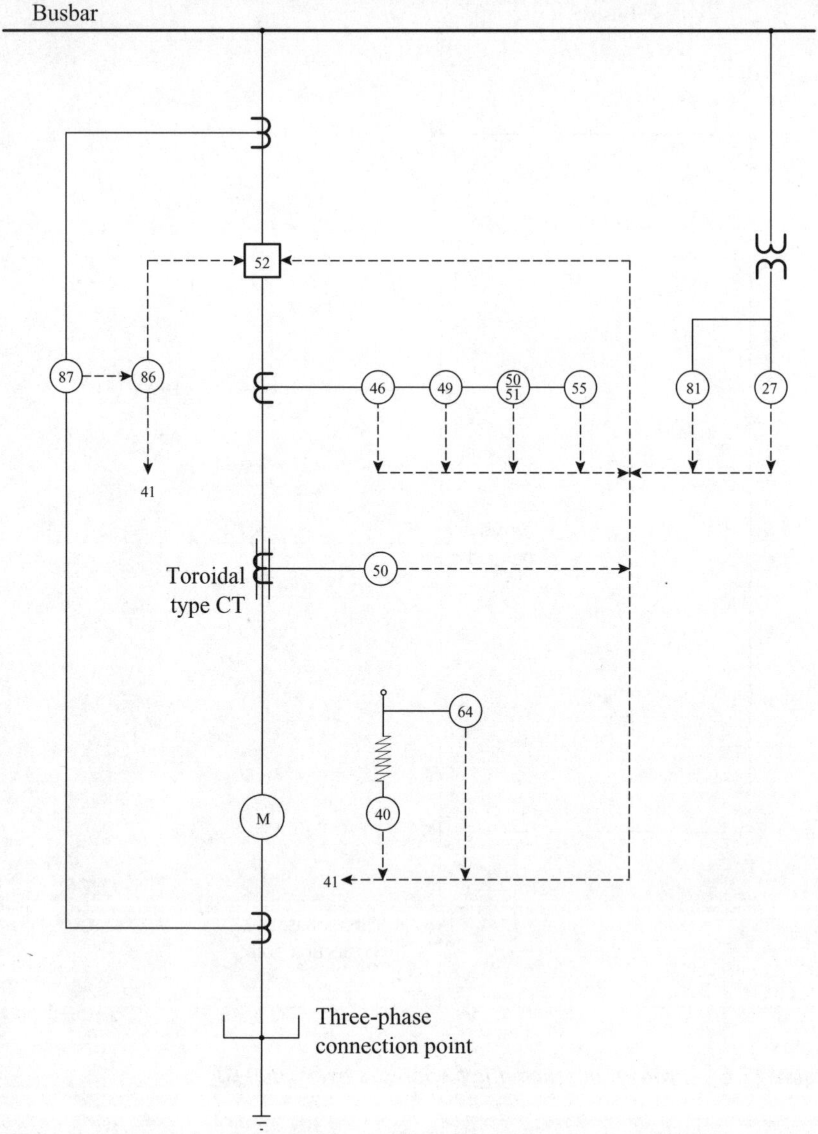

Busbar

Figure 12.7   Protection schematic for synchronous motors over 1000 HP

12.1.3   *Transformer protection*

Transformer protection should take into account the power, voltage, vector group, and the importance of the unit within a particular system. Depending on these factors, the transformers can be assigned to one of the two groups, as below.

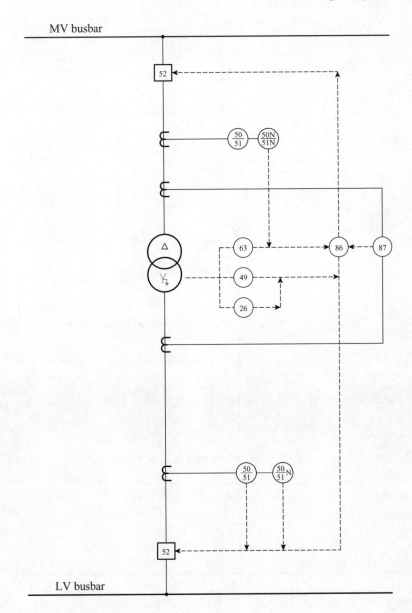

Figure 12.8   Protection for MV/LV transformer

MV/LV transformers

The protection of these units should include overcurrent protection for both the MV and LV windings, plus devices such as over-pressure protection (e.g. Buchholz surge), and thermal protection, as indicated in Figure 12.8. Typical ratios for inter-busbar MV/LV transformers at substations are 33/11, 34.5/13.2, and 13.2/4.16 kV.

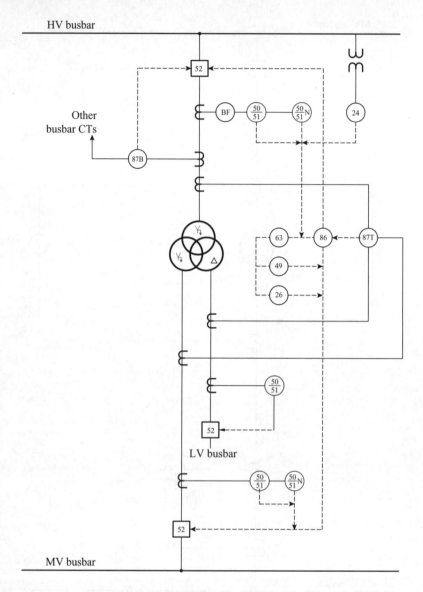

Figure 12.9   Protection schematic for HV/MV/LV transformer

HV/MV/LV transformers and autotransformers

In addition to the protection listed for the MV/LV transformers, the protection for transformers in this group should include overall differential protection, which is essential because of its reliability and high speed of operation. In this case, shown in Figure 12.9, since the transformer has three windings a three terminal type of differential protection is required. The diagram also includes the differential busbar protection, which is usually installed on large transformers connected to HV busbars. As the transformer has an LV winding, overcurrent protection for this winding has been included as well. Typical ratios for HV/MV transformers are 132/33, 145/11, 132/11, 115/34.5 and 115/13.2 kV.

It is common to use autotransformers where large powers are involved and the voltage ratio is around 2 : 1. Typical ratios for autotransformers are 275/132 and 230/115 kV. The protection schemes for autotransformers are very similar to those for HV/MV/LV transformers since autotransformers can be treated as three winding units for protection purposes. The protection for this type of transformer is essentially the same as that quoted in the previous paragraph, taking into account the modifications to the overcurrent relay connections (see Figure 12.9).

12.1.4   Line protection

Line protection generally consists of overcurrent, distance, and directional overcurrent relays and, depending on the voltage level at which the line is operating, schemes indicated below are in general use.

Medium voltage lines

These circuits should be protected with overcurrent relays, and directional overcurrent relays should be used on an MV ring, while MV radial circuits should include reclose relays within the protection scheme. The schematic diagrams for these two cases are shown in Figure 12.10. It is common practice on MV feeders for the tripping of instantaneous relays to be routed via the reclosing relays. This ensures that the reclosing relay is energised from a protection relay trip to start a reclose operation. Time-delay units are often used to produce a definite time trip, i.e. without the possibility of a reclose operation occurring.

High voltage lines

HV lines would normally have distance and directional overcurrent protection, plus carrier wave receive and reclosing relays. Duplication of the main protection can sometimes be justified on important lines, or the use of some other type such as differential protection. When radial circuits are involved, directional overcurrent relays can be replaced by nondirectional overcurrent relays. The schematic diagram for a typical protection arrangement for HV lines is illustrated in Figure 12.11.

12.2   Substation design diagrams

A protection engineer should have a good understanding, not only of the performance of relays and the criteria for setting these as set out in the previous chapters, but also of the relationship between protection and the other equipment in the substations and the distribution system. This section, although not intended to cover substation design, includes some basic information on substation equipment layout, and on other diagrams that a protection engineer should be able to handle without difficulty,

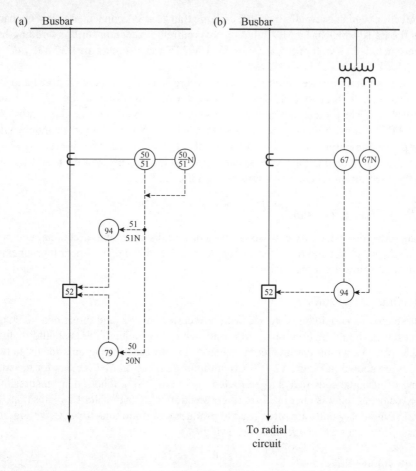

Figure 12.10   Protection arrangements for feeders and MV ring lines: (a) feeder
protection; (b) MV ring protection

in order to ensure a better appreciation of protection schemes and relay settings, and
operational procedures.

Apart from the pure electrical aspects, the design of a substation incorporates
several engineering fields, among them civil, mechanical and electronic. Within
the electrical design function, the basic diagrams used are the single-line diagram,
substation equipment layout drawings, diagrams of AC and DC connections, and secondary wiring plus logic diagrams. A brief mention of these is given in the following
paragraphs.

12.2.1   Single-line diagrams

A single-line diagram shows the disposition of equipment in a substation, or network,
in a simplified manner, using internationally accepted symbols to represent various

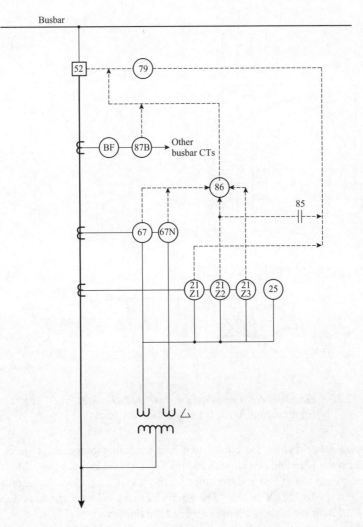

Figure 12.11 Typical protection arrangement for HV lines

items of equipment such as transformers, circuit breakers, disconnectors etc., generally with a single line being used to represent three-phase connections. Often the main data for the HV equipment is included on the diagram. More detailed single-line diagrams include such items as the instrument transformers, and the protection, measurement and control equipment and their associated secondary wiring.

12.2.2 Substation layout diagrams

Substation layout diagrams provide scale drawings of the location of each piece of equipment in a substation, both in plan and elevation. While individual utilities may have their own format, there is a high degree of standardisation of these types

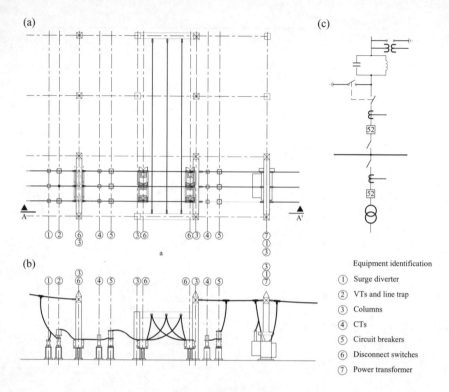

Figure 12.12   *General arrangement for two 115 kV bays:* (a) general layout;
(b) elevation A-A′; (c) single-line diagram

Equipment identification

① Surge diverter
② VTs and line trap
③ Columns
④ CTs
⑤ Circuit breakers
⑥ Disconnect switches
⑦ Power transformer

of drawings worldwide, for contractual and tendering purposes. Figures 12.12a
and b show the plan and elevation drawings for a typical layout of two 115 kV bays,
one for a transmission line and the other for the HV side of a local transformer, connected to a single 115 kV busbar. The equivalent single-line diagram is depicted in
Figure 12.12c at the top right-hand corner of the drawing.

Although protection engineers may not be directly involved with layout diagrams,
these drawings do show the relationship between various items of primary equipment
and the location of those items associated with protection systems – for example,
current and voltage transformers that may be located separately from other items
of equipment or placed within high voltage equipment such as circuit breakers. The
protection engineer is thus able to ensure that he can safely locate protection equipment
within the substation.

12.2.3   Diagrams of AC connections

A diagram of AC connections generally shows the three-phase arrangement of the
substation power equipment, and the AC circuits associated with the measurement,
control and protection equipment, in schematic form. The AC diagrams for a typical

substation contain information corresponding to bays for incoming transmission lines, bus section and bus couplers, power transformers, and MV feeder circuits. In addition there would also be diagrams containing information on such items as motors and heating that operate on AC.

The layout of AC connections diagrams should be carried out observing the following points. Each diagram should include all equipment corresponding to a bay, with breakers, disconnectors and transformers represented by schematic symbols. In CT current circuits only the current coils of the measurement instruments and the protection relays should be drawn, indicating clearly which coils are connected to each phase and which to the neutral. The polarity of equipment should be indicated on the drawings. It is useful to indicate equipment whose future installation can be foreseen by means of dotted lines.

Solid-state relays should be represented schematically by squares, showing the number of terminals and the method of connecting the wiring carrying the voltage and current signals. The points where a connection to earth exists should also be indicated in this diagram, for example when the neutral of the measurement transformers is connected in star.

The main nominal characteristics should be marked close to each item of equipment. For example, for power transformers, the voltage ratio, power rating, and vector group should be provided; for power circuit breakers, the nominal and short-circuit current ratings; the transformation ratios for voltage and current transformers, and the nominal voltage of lightning arresters. Voltage transformer circuits should be drawn physically separated from the rest of the circuits, and the connections to the coils of the instruments that require a voltage signal should also be indicated. As a minimum, the AC diagram of a transformer should include all the equipment in the bay between the high voltage busbar and the secondary bushings of the transformer.

12.2.4   Diagrams of DC connections

Diagrams of DC connections illustrate the direct current circuits in a substation and should clearly show the various connections to the DC auxiliary services. These diagrams contain information corresponding to equipment such as breakers and disconnectors, protection and control systems for transformers, busbars, transmission lines and feeders, annunciator systems, motor and heating circuits that operate on DC, and emergency lighting and sockets. A diagram of connections for all substation equipment that takes supplies from the DC system should be provided.

The positive infeeds are normally shown at the top of the diagram, and the negative ones at the bottom and, as far as possible, the equipment included in the diagrams should be drawn between the positive and negative busbars. Due to the considerable amount of protection and control equipment within a substation it is generally convenient to separate out the DC connections into different functional groups such as control and protection equipment, and other circuits such as motors, heating etc.

It is common practice to draw dotted horizontal lines to indicate the demarcation between the equipment located in the switchgear and that located in the protection relay panels. It is useful if the signalling and control equipment in the relay and control panel is located in one part of the diagram, and the protection equipment in another part. Every terminal should be uniquely identified on the drawing. As far as possible, the contacts, coils, pushbuttons and switches of each mechanism should be drawn together and marked by a dotted rectangle so that it is easy to identify the associated equipment and its role in the circuit.

The internal circuits of the protection equipment are not shown, since it is sufficient to indicate the tripping contacts and the points of interconnection with other equipment inside a dotted rectangle. Given the complexity of distance relays, it might be necessary to make a separate diagram to indicate their connections to the DC system and the interconnection of the terminals. It is also possible that separate diagrams may be required for transformer and busbar differential protection.

Each power equipment bay should have two DC circuits; one for feeding the protection equipment and a separate one for signalling purposes and controlling breakers and disconnectors. The two supplies should be kept independent of each other and care should be taken to avoid connecting any equipment across the two DC supplies.

12.2.5   Wiring diagrams

Wiring diagrams show the interconnection of the multicore cables, for example between the switchgear and the associated control panels, and the routeing of individual wires to the equipment installed in the relay and control panels. These diagrams are required to facilitate the wiring of the measurement, protection and control equipment at the substation construction stage. The wiring should be carried out in accordance with the layout shown in the AC and DC diagrams.

It is logical that the layout of the different devices on the wiring diagrams should be as seen from the rear of the relay and control panels, as in practice. Each device should be represented by its schematic, with every terminal located in accordance with its actual position on the panel. Each conductor should be marked with the same identification code as the terminal to which it is connected, and also marked at each end with the location of the far end of the conductor, according to a predetermined code. To make the wiring easier to install, the location of the wires on the wiring diagram should correspond to their proposed location inside the relay and control panel. In the wiring diagrams the following elements should be uniquely identified – terminals and sets of terminals; multicore cables that go to the switchgear; conductors that go from individual terminals to equipment located in the relay and control panels; equipment installed in the relay and control panels.

Multicore cables

Each multicore cable should have an identification number; in addition every conductor in each cable should be numbered. It is useful if the numbering of multicore cables is carried out consecutively by voltage level. With this in mind, an ample range of numbers should be provided, for example multiples of 100 for each voltage level, thus ensuring that there are sufficient spare consecutive numbers available for any additional cabling in the future. All the conductors in the wiring diagram should be marked at each end with the location of the far end of the conductor.

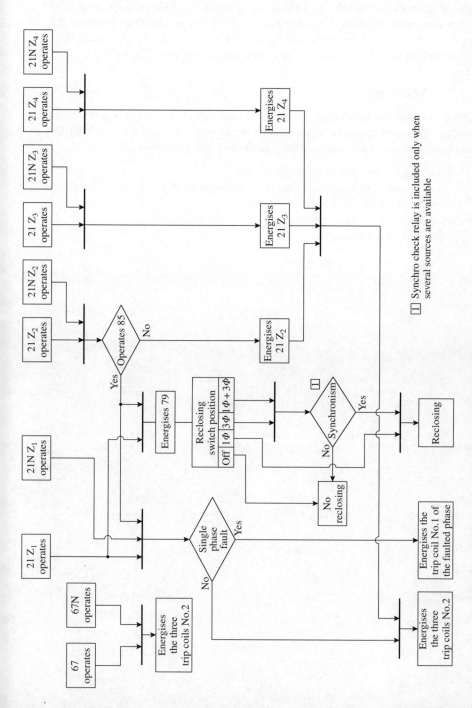

Figure 12.13 Protection logic schematic for 115 kV line bay

12.2.6   Logic diagrams

These diagrams represent the protection schemes for the different substation bays by means of normalised logic structures in order to show in a structured manner the behaviour of the substation protection system for any contingency. An example of such a diagram for a 115 kV line bay at a substation is shown in Figure 12.13.

12.2.7   Cabling lists

Cabling lists provide information on the multicore cables that run between various items of equipment and help to make it easier to verify the substation wiring for maintenance work. The lists should include the following information:

- number, length and type of multicore cable;
- colour or number of each conductor in the multicore cable;
- identification of each end of the conductor;
- identification of the equipment at each end of the conductor;
- the function of the conductor.

Chapter 13

Processing alarms

13.1 General

The various protection schemes described in earlier chapters in this book are provided to disconnect faults on power systems promptly, with the minimum of disruption to the network and to customers. Alarms are triggered by the operation of the protection relays and send coded information to distribution control centres so that the control room operators are aware of what is happening on the network. Other alarms indicate the state of the power system, for example voltages at various locations and load flows on the more important circuits.

These alarms provide one of the main sources of information flowing in real-time into a distribution control centre and are normally channelled to one printer in the control room where a hard copy can be produced. The alarm streams are also channelled to the operator's or control engineer's console where they can be displayed on computer screens. A third avenue for alarm streams is for them to be stored in a data logger where the principle function is to retain a history of the alarm streams, which can be used for post-system fault analysis if a serious fault occurs in a system and it is thought necessary to carry out such an analysis (see Figure 13.1).

It is worth nothing that, with this arrangement, the alarms are not processed. Several events may occur simultaneously, or at a close time proximity with respect to each other, and each incident may trigger many alarms resulting in a large number of alarms flowing into the control centre in close succession. The operator would then have to use his/her judgement based on experience to decide what exactly has happened to the system. Subsequent telephone calls from customers may also help to determine the exact location of these incidents such as a blackout in a certain district. The aim of an alarms processor is to help the operator to arrive at a sensible conclusion speedily and to discard redundant information in the alarm streams.

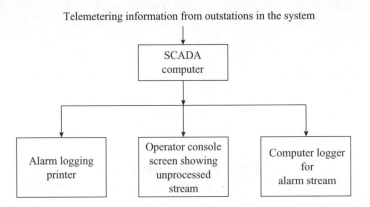

Figure 13.1   Alarm streaming

An alarm is a signal that may be analogue or digital in nature and each alarm carries certain information, which might include for example:

- notification of the operation of protective equipment including tripping, reclosing and warning messages;
- indication of the change of state of a particular piece of equipment – e.g. a circuit breaker opening or closing;
- warning of potential danger with plant, lines or cables unless remedial action is taken – e.g. cable low oil pressure or tripping battery low;
- high or low voltage levels and loads on plant items.

Different utilities or power companies may use various protocols to identify their alarm messages – for example the letter 'N' could indicate a specific type of alarm. In addition the priority level of the alarm can be shown, if required. It is also necessary to be able to indicate the time of day, and the name of the substation where the alarm has originated. The type of alarm also needs to be identified, e.g. 'AUTOTRIP' can mean automatic trip. Then codes such as 'OPRT*' may be used to indicate that some equipment has operated, with the * indicating the alarm has automatically reset itself. Some messages are quite brief and give little information beyond, say, the fact that at a particular time an automatic trip has operated at a named substation, and that the alarm has reset itself. There is no information about which trip has operated, and whether it has gone from an open state to a closed state or vice versa. Other types of alarm may give more information. For example I 1729 JOHNSTON 33 BSECT OPEN could be used to indicate that the 33 kV busbar section circuit breaker at Johnston substation has opened at 1729 hours.

13.2   Alarm processing methods

Two simple methods of processing alarms are used widely in many existing control centres. In one, all the alarms are assigned a certain level of fixed priorities and are

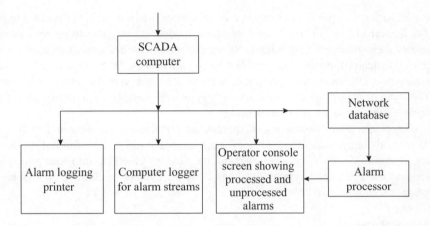

Figure 13.2   Extended alarm processor

then displayed to the operator once they have been activated in their pre-fixed priority level. The alarms may also have a fixed route to a specific console, which is often related more to the geographical location of the alarm than to priority levels. The second method is more flexible than the first. Under normal conditions, all levels of alarms will be displayed. However, under emergency conditions, the operator has the option to suppress certain levels of alarms coming to the console to avoid being overloaded with incoming data. This enables the operator to be less stressed and more able to cope with the incoming information and reach a decision on system conditions.

An additional path can be added to the control room alarm system, comprising a network database and an alarms processor (see Figure 13.2). The former contains the original network topology and also real-time updated topology. It also contains updated information on each substation including the rating and voltage levels of transformers, busbar couplers, protection equipment, auto-reclosure equipment, fault throwing switches and other information that the alarms processor unit takes into consideration when interpreting alarm messages from the system. The network topology information is updated in real-time while other information may be updated in an off-line mode. This information, together with the alarm streams, is fed into the alarms processor which will automatically identify the actual events on the system. The processed alarms, together with the unprocessed ones, will be on display to the control room operator. The system acts as an aid to the operator who will then have to decide whether the processor has arrived at a sensible conclusion about the events.

13.3   Expert systems

The term expert system is used in reference to a computer program that is assumed to have the same knowledge as a human expert, and which will attempt to simulate his/her reasoning process in order to arrive at a solution to a given problem or a given set of conditions. In its simplest form the program will be structured in the form of a tree, with each option representing a node with only two branches connected to it. The answer of 'yes' or 'no' to each option will allow the program to proceed along one of these branches that leads to the next option. The decision to go along one branch instead of the other is governed by so-called rules. In this case it is a simple yes/no rule. Ultimately the system should arrive at a solution of the network problem. This type of expert system is relatively easy to construct, but it is not suitable for complex tasks such as alarm processing.

There are many techniques or approaches for alarm processing and each has its own advantages and disadvantages. Each approach has its own form of presenting alarms to the operator and in general they can be classified into four principle approaches. The expert system should embody the essence of all these approaches, which are summarised below.

Alarm processing

This is an important approach for a large control room that covers a wide geographical area, and where each operator is responsible for a specific region within this area. Alarms in a particular region are only presented to the relevant operator's console. However, top-level alarms that are considered to have grave severity may also be diverted and displayed simultaneously at the senior operator's console.

Alarm prioritisation

In this approach, alarm messages are displayed in a priority order list on the operator's console. This approach may be in addition to the alarm processing referred to in the paragraph above.

Alarm filtering

This approach aims to reduce the number of alarms each operator receives. It eliminates and suppresses alarms that simply duplicate the information from alarms received earlier. Finally, alarm reduction would be possible by suppressing all alarms that are expected, for example, as a result of an operator's actions or a correct protection system operation.

Alarm summarising

In this approach, alarms relating to one perceived event are replaced by a simple summary message stating the possible cause of all these alarms.

13.4　Equivalent alarms

Alarm systems in an electricity utility are often not installed at the same time, but as required as the electrical network develops. In addition, they may have been installed by different manufacturers using different parts, so that the alarms that arrive at the control room for the same type of event may be constructed differently. For example, the message 'alarm suppressed' may be displayed in a number of different ways, e.g. ALM.SUPP, ALMS SUP, ALM SUPP, ALRM SUP, etc. Examples of other equivalent alarms are IT RECV or I/T REC for intertrip signal received, 33AR ALM, 33KV A/R, AR ALM33 for an alarm originating from a 33 kV substation, and CP LO, LOP, LO OIL P, CABL LOP for a cable low oil pressure alarm.

13.5 Rules

The reasoning mechanism of an expert system is via the rules set up within the program. Some of these rules are illustrated below:

- event – successful closure of {CB} under operator control, where {CB} is the description of a given circuit breaker in a substation; expect that an operator closes the circuit breaker – action OpAction Close {CB} – and that the indication CLOSED {CB} is received;
- event – volts high on main busbar 1 at {StationName}; expect to receive alarm MN.B/B.1 HIGH OPERATED {StationName};
- event – {StationName}; voltage on main busbar 1 returned to normal; expect to receive alarm MN.B/B.1 HIGH_RESET {Station Name}.

The first alarm shown above is an event that indicates a successful closure of a specific circuit breaker under operator instruction. In order for this event to be true, two actions are needed. The first one is the operator action leading to the closing of the {CB}, and the second one is the alarm that comes back from the substation indicating that the {CB} has closed. The rules are simple but the different combinations and the overlaying of these rules can be used to reason out complex situations. Thus equivalent messages allow the system to interpret different descriptions of alarms that have the same meaning, while rules are used to interpret the combined meaning of alarms.

13.6 Finger printing approach

One approach that may be used to handle situations of wide variety and complexity is the finger printing approach which, as the name suggests, involves matching current alarms with the set of alarms that would be expected for a particular type of system event. If there is an exact match then it is concluded that this event has occurred.

Consider the network in Figure 13.3 where, because of the indicated fault, the line between substations A and C has tripped out. The fault will generate a series of alarms as shown on Table 13.1.

The first alarm indicates that an automatic trip has operated in substation A while the second alarm indicates that circuit breaker 5 in substation A has opened. Similarly the third alarm indicates that an automatic trip has operated in substation C with the subsequent two alarms indicating that circuit breakers 2 and 3 at substation C have also opened. In a finger printing approach, such an alarm list is stored in the database

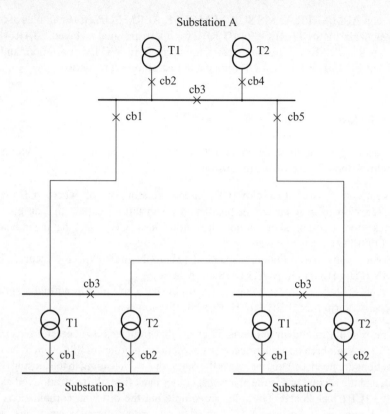

Figure 13.3   Network example

Table 13.1   Alarms for network in Figure 13.3

Time	Substation	Circuit	Action
1033	SUBSTN A		AUTOTRIP OPRT*
1033	SUBSTN A	CB5	OPEN
1033	SUBSTN C		AUTOTRIP OPRT*
1033	SUBSTN C	CB3	OPEN
1033	SUBSTN C	CB2	OPEN

and, if an exact match occurs, then it can be concluded that there has been a fault in line AC.

The finger printing approach works well under an ideal situation. However, in an alarms processing situation, it is not the best choice. This is because it is unable to cope with missing alarms or a piecemeal arrival rate of alarms, or to relate to different system events occurring simultaneously.

13.7 Hypothesis approach

An alternative to the fingerprinting method is the hypothesis approach. In this method a database is set up identifying those alarms that should be received at the control centre for any incident on the power system. As alarms arrive, a series of hypotheses that might account for the alarms received so far can be generated and these hypotheses can later be confirmed or discarded as more alarms arrive progressively at the control centre. Figure 13.4 provides a simplified schematic diagram of the flow chart for a hypothesis expert system developed at the Department of Electronic and Electrical Engineering, University of Strathclyde, Glasgow, UK.

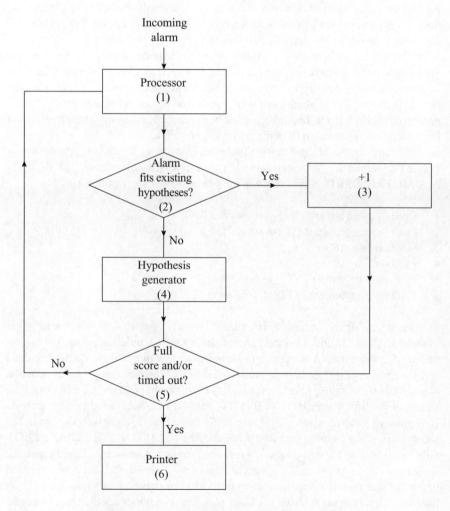

Figure 13.4   Flow chart for a hypothesis expert system

When an alarm is fed into the input processing section of the expert system (1), it is broken down into separate fields and converted for ease of manipulation. The next stage is to check if the incoming alarm fits into any of the existing hypotheses (2). If the result is positive, the point scores of those hypotheses containing this alarm are increased (3) and the results are passed on to the evaluation stage (5) where each hypothesis is checked to see if any have reached a full points score, and/or reached the pre-determined time that could be one or two polling periods, depending on the nature of the telemetering system.

If a full points score has been achieved then the program concludes that a solution has been found and prints out a summary message on the operator's console (6). If any hypothesis has timed out without all the expected alarms having arrived then the most likely hypotheses are printed out, indicating those alarms that were expected but had not been received. If a hypothesis has not reached a full score nor timed out, then this is recycled back to the input section to repeat the checking process to see if any other relevant alarms have been received.

If the incoming alarm does not fit into any of the hypotheses stored in the program then new hypotheses have to be generated (4). It may be necessary to expand the rules to form completely new hypotheses, and produce listings of expected alarms for each new hypothesis. These alarms are then transferred to the evaluation stage (5) and processed to check if a full points score has been reached and/or any of the hypotheses have timed out, as referred to in the previous paragraph.

Referring to the alarms list in Table 13.1, relating to the system shown in Figure 13.3, under the hypothesis approach the arrival of the first alarm 1033 SUBSTN A AUTOTRIP OPRT* would result in the generation of six hypotheses:

1. Fault in transformer 1 (T1) at substation A.
2. Fault in transformer 2 (T2) at substation A.
3. Fault on line AB.
4. Fault on line AC.
5. Fault on transformer 1 (T1) at substation B.
6. Fault on transformer 2 (T2) at substation C.

The arrival of the second alarm 1033 SUBSTN A CB5 OPEN helps to reject hypotheses 1, 2, 3, and 5, since none of these faults would have required circuit breaker 5 at substation A to open in order to isolate any one of these faults. The third and fourth alarms do not help to eliminate either of the remaining two hypotheses (4 and 6). However, they help to confirm that these two hypotheses are still valid. The arrival of the fifth alarm, 1033 SUBSTN C CB2 OPEN, does not alter the situation. Hypotheses 4 and 6 are still valid. However, for hypothesis 6, (fault on transformer 2 at substation C), one further alarm would be expected – SUBSTN C TR2 TRANS PROT OPRT – indicating that the protection for transformer 2 at substation C had operated. Under such circumstances, the alarms processor may have to wait for a given period before it can arrive at a decision. As no further alarm arrives, hypothesis 4 – fault on line AC – becomes the selected solution. This example illustrates the basic principle of the hypothesis approach, but it also raises several important points, such as how

long the alarms processor should wait before arriving at a decision, how to deal with
missing alarms and with simultaneous events.

The following examples demonstrate different capabilities of the expert system
and the inference mechanism behind the system.

Example 13.1

The first example covers the substations at Iverley, Norton and Pedmore in the network
shown diagrammatically in Figure 13.5, where circuit breakers marked with a circle
operate in a normally open position.

The alarms received for this event are shown in Table 13.2.

From Table 13.2, the first alarm for this event arrives at 0821 stating 0821
IVERLEY AUTOTRIP OPRT*. The alarms processor recognises this as meaning that
there has been an automatic trip of one or more circuit breakers at Iverley substation

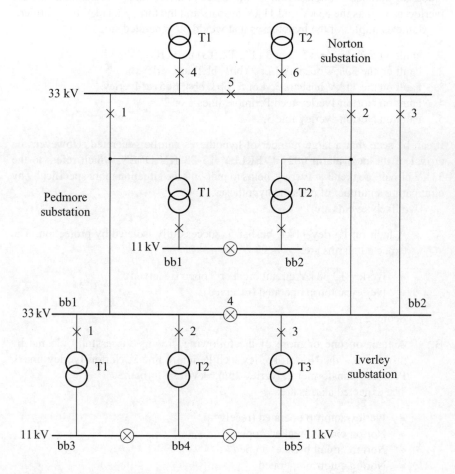

Figure 13.5   Network for Example 13.1

Table 13.2   Alarms received for Example 13.1

Time	Substation	Circuit	Action
0821	IVERLEY		AUTOTRIP OPRT*
0821	IVERLEY	T3 33 KV CB	OPEN
0821	NORTON		AUTOTRIP OPRT*
0821	NORTON	NRTN-IVLY/PED 2	OPEN
0821	NORTON	NRTN-IVLY 3	OPEN
0822	PEDMORE	T2 11 KV CB	OPEN

due to protection operation. As only one alarm has been received so far, the alarms processor initiates hypotheses for a fault on each of the three circuits connected to Iverley, as well as the 33 kV and 11 kV busbars and the three 33/11 kV transformers.

Some examples of the hypotheses that would be generated are:

1. Fault on transformers 1, 2 or 3 (T1, T2, T3) at Iverley.
2. Fault on the 33 kV busbars 1 or 2 (bb1, bb2) at Iverley.
3. Fault on the 11 kV busbars 3, 4 or 5 (bb3, bb4, bb5) at Iverley.
4. Fault on Norton-Iverley teed Pedmore lines 1 or 2.
5. Fault on Norton-Iverley line 3.

It can be seen that a large number of hypotheses can be generated. However, the arrival of the next alarm 0821 IVERLEY T3 33 KV OPEN, which refers to the 33 kV circuit breaker 3 at Iverley, helps to pinpoint the situation more specifically by eliminating a number of possible hypotheses.

Two likely events are:

A.   A fault on Iverley 11 kV busbar 5, successfully isolated by protection. The expected alarms are:

   • Iverley T3 33 kV circuit breaker 3 open (received).
   • Iverley autotrip operated (received).

Score: 2

B.   A fault on one or more of the following circuits successfully cleared by protection – the Norton-Iverley teed Pedmore line 2, Norton-Iverley line 3, Pedmore transformer 2, Iverley 33/6.6 kV transformer 3.
The expected alarms are:

   • Iverley autotrip operated (received).
   • Norton circuit breaker 2 open.
   • Norton circuit breaker 3 open.
   • Norton autotrip operated.
   • Iverley 33 kV circuit breaker 3 open (received).

- Pedmore transformer 2 11 kV circuit breaker open.
- Pedmore autotrip operated.

Score: 2

The next alarm received – 0821 NORTON AUTOTRIP OPRT* – is expected if hypothesis B is correct, so the score for this hypothesis is increased by a further point to 3. The following three alarms also contribute to hypothesis B, which finally reaches a score of 6.

One expected alarm – PEDMORE AUTOTRIP OPRT* – has not been received. However, this does not invalidate hypothesis B as being the preferred solution and the alarms processor prints out hypothesis B as the final solution. This would include a list of nodes describing the fault area since it is not possible to be more specific about the location of the fault. The network bounded by the Norton-Iverley line 3, the Norton-Iverley teed Pedmore line 2, transformer 2 at Pedmore, busbar 2 at Iverley, and transformer 3 at Iverley is defined as the blackout area.

Figure 13.6 shows the single line diagram for the next two examples. The first one is centred around Enville substation, and the second one around Clent substation.

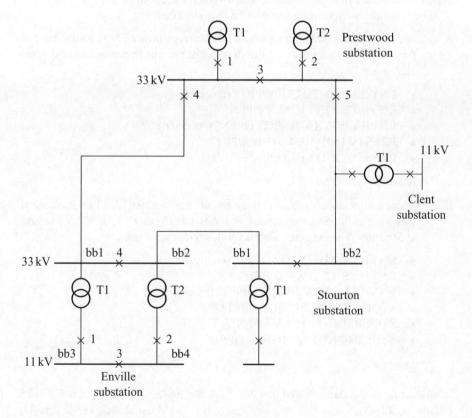

Figure 13.6   Network for Examples 13.2 and 13.3

It should be noted that those circuit breakers marked with a circle normally operate in the open position.

Example 13.2

Upon the arrival of the first alarm of this event – 1425 ENVILLE AUTOTRIP OPRT* – the alarms processor initiated a number of hypotheses to cover all the possible fault locations:

1. Fault on transformer 1 (T1) at Enville, 33 kV busbar 1 (bb1) at Enville, and the Prestwood-Enville circuit.
2. Fault on transformer 2 (T2) at Enville, 33 kV busbar 2 (bb2) at Enville, the Enville-Stourton circuit, transformer 1 (T1) at Stourton.
3. Fault on 11 kV busbar 3 (bb3) at Stourton.
4. Fault on 11 kV busbar 4 (bb4) at Stourton.

The arrival of the second alarm – 1425 ENVILLE 33 KV B.SECT OPEN* – indicates that the 33 kV bus-section circuit breaker at Enville was opened and then reclosed. With these two alarms having been received, which are relevant to the first two hypotheses, these two hypotheses are thus awarded a score of 2.

The most likely hypotheses can now be narrowed down to:

A. Successful reclosure after a fault on Enville transformer T1, 33 kV busbar 1 at Enville, or on the Prestwood-Enville circuit. For this hypothesis the expected alarms are:

    • ENVILLE AUTOTRIP OPRT* (received).
    • ENVILLE T1 11 KV CB OPEN*.
    • ENVILLE 33 KV B.SECT OPEN* (received).
    • PRESTWOOD AUTOTRIP OPRT*.
    • PRESTWOOD CB4 OPEN*.

    Score: 2

B. Successful reclosure after a fault on Enville transformer T2, 33 kV busbar 2 at Enville, Enville-Stourton circuit, Stourton transformer T1, or 33 kV busbar1 at Stourton. The expected alarms for this hypothesis are:

    • ENVILLE AUTOTRIP OPRT* (received).
    • ENVILLE T2 11 KV OPEN*.
    • ENVILLE 33 KV BSECT OPEN* (received).
    • STOURTON AUTOTRIP OPRT*.
    • STOURTON T1 11 KV OPEN*.
    • STOURTON 33 KV BSECT OPEN*.

    Score: 2

Referring to the alarms listed in Table 13.3, the arrival of the third alarm – 1425 ENVILLE T1 11KV CB OPEN* – relates to T1 11 kV circuit breaker at Enville, which has opened and successfully reclosed. Therefore the score for hypothesis A

Table 13.3   List of alarms received for Example 13.2

Time	Substation	Circuit	Action
1425	ENVILLE		AUTOTRIP OPRT*
1425	ENVILLE	33KV B.SECT	OPEN*
1425	ENVILLE	T1 11KV CB	OPEN*
1425	PRESTWOOD	PRSTWD-ENVIL	OPEN*

Table 13.4   Alarms for Example 13.3

Time	Substation	Circuit	Action
1748	CLENT	T1 33 KV CB	OPEN
1748	CLENT	T1 33 KV CB	O/S
1750	CLENT	T1 11 KV CB	O/S

has its score increased by 1 to 3, whereas hypothesis B still has a score of 2, and the other hypotheses are no longer viable.

The fourth alarm – 1425 PRESTWOOD PRSTWD-ENVIL OPEN* – indicates that circuit breaker 4 at Prestwood, controlling the circuit to Enville, has opened and subsequently reclosed. Since this message is expected for hypothesis A, its score is increased incrementally once more to a value of 4. This leaves hypothesis A short of one alarm – PRESTWOOD AUTOTRIP OPRT*. However, since the alarm PREST-WOOD CB4 OPEN* has been received, it is not necessary to wait for this final alarm before being able to arrive at a decision. If no more alarms relating to these two hypotheses arrive within the next polling period then the alarms processor has to weigh up these hypotheses and make a decision as to the most likely one. On this case, hypothesis A has the highest score and is selected as the preferred event.

Example 13.3

For this event the alarms are shown in Table 13.4.

The first alarm indicates that the 33 kV circuit breaker at Clent substation has opened. Such a message could indicate either of the following likely possibilities.

1. The circuit breaker has opened due to an autotrip initiated by protection operation. Therefore, further alarms would be expected together with appropriate autotrip alarms.
2. The circuit breaker has been opened prior to maintenance work being carried out on associated equipment.

The next two alarms add weight to the probability that some form of outage is being performed, with two hypotheses being presented to the operator.

A.   Fault on Prestwood-Stourton teed Clent circuit, Clent transformer 1, Stourton 33 kV busbar 2. For this hypothesis the expected alarms are:

- CLENT T1 33 KV CB OPEN (received).
- PRESTWOOD CB5 OPEN.
- PRESTWOOD AUTOTRIP OPRT*.
- STOURTON AUTOTRIP OPRT*.
- STOURTON 33 KV CB OPEN.
- CLENT T1 11 KV CB OPEN (received).
- CLENT AUTOTRIP OPRT*.

This results in a score of two alarms received out of a possible seven.

B.   Suspected maintenance on Clent transformer 1, Prestwood-Stourton teed Clent circuit, and Stourton 33 kV busbar 2. The expected alarms for this hypothesis are:

- PRESTWOOD CB5 OPEN.
- PRESTWOOD CB5 O/S.
- STOURTON CB2 OPEN.
- STOURTON CB2 O/S.
- CLENT T1 33 KV CB OPEN (received).
- CLENT T1 33 KV CB O/S (received).
- CLENT T1 11 KV CB OPEN (received).
- CLENT T1 11 KV CB O/S.

Score: 3

The scores for both hypotheses are not high, but the fact that no AUTOTRIP alarms have arrived strengthens the hypothesis that the event is due to maintenance work rather than a fault event, which should have initiated protection operation. After waiting for a further polling period to ensure that no more alarms related to this event have arrived, all the competing hypotheses are weighed up, and the solution of hypothesis B is chosen.

The three examples have illustrated three different conditions. The first one had a full score and the system identified a blackout area with the names of the nodes identifying that area. The second example illustrates the ability of the system to cope with missing alarms, and detected a successful reclosure after a fault in a given area identified by a set of node names. The third example shows how the alarms processor copes when there are substantial missing alarms and the system can only conclude a suspected situation.

An alarms processor programmed to operate on a hypothesis approach basis can cope with a large number of incoming alarms from different network incidents, even if these are time-coincidental, for example during a thunderstorm with multiple lightning strikes on the power system. The ability of the process to assemble the appropriate alarms together for each incident, and to reach a decision on each incident, provides a powerful aid to the system control room operator when the alarm load is most demanding and is also a most useful analytical tool for subsequent fault analyses.

Chapter 14

Installation, testing and maintenance of protection systems

Although the aim of this book is to provide the basis to guarantee a suitable relay setting procedure in distribution networks, it is felt that some reference should be made to the installation, testing and maintenance of protection systems. No matter how well the relay applications are carried out, a protection scheme is worthless if its actual performance cannot be guaranteed. It is important to emphasise that a protection scheme covers not only the relays but also the CTs and VTs that feed them, and the circuit breakers that open the circuits on receipt of a trip signal from the relays when a fault occurs.

14.1 Installation of protection equipment

The installation of protection relays should be carried out following the instructions normally included by manufacturers in their service manuals. The relays are installed on control panels that should preferably be situated in areas free of dust, damp, excessive vibration, and heat. The older types of relays may be set into the panel (flush fitted) or mounted on the face of the panel with a suitable casing. Multifunction relays are generally mounted on standard 19-inch rack cases. The heights of the cases are specified in terms of the number of standard rack units, which typically range from 2 to 5 rack units. The depth of rack mount cases varies depending on the relay model. Some relay models include built-in facilities for testing the relay in the case. Other relay models require external provisions to test the relay in its case. Figure 14.1 shows typical layouts of protection and control panels for HV and MV substation bays. The dimensions correspond to typical equipment from various manufacturers.

The electrical connections to the relay terminal should, as far as possible, be made with flexible stranded copper wiring, and the terminals should be designed so that a solid connection with effectively no resistance can be guaranteed. Great care must be taken to ensure that the CT secondary wiring circuits are not open-circuited when primary current is flowing since this could cause damage to the relays, wiring and the

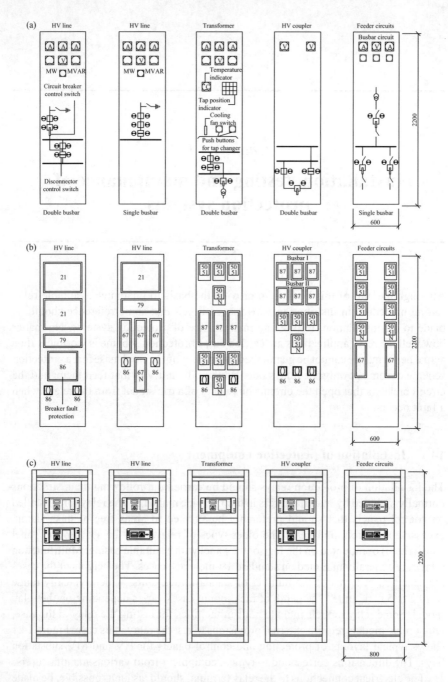

Figure 14.1   Layout of equipment on control panels: (a) front view of panel; (b) rear view of panel with electromagnetic relays; (c) rear view of panel with numerical relays

CTs due to the high voltage that can be generated. Before starting to test the relays, any mechanical fastening that the manufacturer has placed on the moving parts to prevent damage during transportation and installation should be removed.

14.2   Testing protection schemes

Testing protection equipment can be divided into three types:

- factory tests;
- pre-commissioning tests;
- periodic maintenance tests.

14.2.1   Factory tests

It is the responsibility of the relay manufacturer to carry out suitable tests on all protection equipment before it is delivered and put into service. Since the protection relays are required to operate correctly under abnormal system conditions, it is essential that their operation be guaranteed under such conditions. The simulation of operational conditions for test purposes is normally carried out during the manufacture and certification of the equipment. These tests should be the most rigorous possible in order to ensure that the protection will operate correctly after transportation to, and installation at, the substation site.

The factory operating tests can be divided into two main groups – those tests in which the relay parameters are reproduced, and those involving the simulation of conditions such as temperature, vibration, mechanical shock, electric impulse, etc., which could affect the correct operation of the relay. In some cases, tests from both groups can be carried out simultaneously.

14.2.2   Precommissioning tests

The most important precommissioning tests and in-service checks can be summarised as below:

- analysis of the wiring diagrams to confirm the polarity of connections, positive- and negative-sequence rotation, etc.;
- a general inspection of the equipment, physically verifying all the connections, at both the relay and panel terminations;
- measurement of the insulation resistance of the protection equipment;
- inspection and secondary injection testing of the relays;
- testing current transformers;
- checking the operation of the protection tripping, and alarm circuits.

The precommissioning tests need to be carefully programmed so that they take place in a logical and efficient order, in order that no equipment is disturbed again during subsequent tests. Before starting the tests it is essential to ensure that the assembly of the particular item being tested has been completed and checked. In addition, the list of the tests to be carried out should be arranged in a chronological order together with any precautions that need to be taken into account. Some of the more usual tests are briefly described below.

Insulation resistance measurement

This test should be carried out with the aid of a 1000 volt Megger. It is difficult to be precise as to the value of resistance that should be obtained. The climate can affect the results – a humid day tends to give lower values, whereas on a dry day much higher values may be obtained.

Secondary injection tests

These tests are intended to reproduce the operating conditions for each relay and are limited solely to the protection as such, so that it is important to read and understand the relay instruction manual (application, operation, technical characteristics, installation and maintenance). In order to carry out these tests it is necessary to electrically isolate the relay by means of test plugs or physically withdraw the relay from its case. Although the relays should have been carefully tested in the manufacturer's works, it is necessary to make some checks on site after they have been mounted on the panels in order to be sure that they have not been damaged in transit to the installation. The actual tests carried out on the relays depend largely on the type of relay.

Secondary injection tests are required to ensure that the protective relay equipment is operating in accordance with its preset settings. Relay inputs and outputs must be disconnected prior to performing these tests. The test equipment supplies the relay with current and voltage inputs that correspond to different faults and different operating situations. Pick-up values are reached by gradually changing the magnitudes of these inputs while simultaneously measuring the relay operating time. Tripping contacts and targets must be monitored during these tests in order to ensure that the relay is working according to the manufacturer's specifications and the settings that have been implemented. If the curves and characteristics of a relay are to be tested at many points or angles, it is convenient to use test equipment that can conduct a test automatically.

Modern protective relay test equipment has the option of performing automatic tests aided by software programs, for which the testing process is much faster and more precise. In addition, the time during which the relay is out of operation is minimised. Figure 14.2 shows a typical layout of a protective relay test equipment. This equipment is able to provide current and voltage injection as well as phase shift when testing directional protection. It thus permits testing of a wide variety of relay types such as overcurrent, directional overcurrent, reverse power, distance and under/overvoltage units.

It is very important to record all the test results, preferably on special forms for each type of relay. For example, a typical pro-forma for overcurrent relay tests, as shown in Figure 14.3, could have the following information recorded:

- basic data about the circuit supplying the relay;
- the settings used before any tests commenced, which had been applied in accordance with the protection co-ordination study. This information should include the pick-up current, time dial, and instantaneous settings;

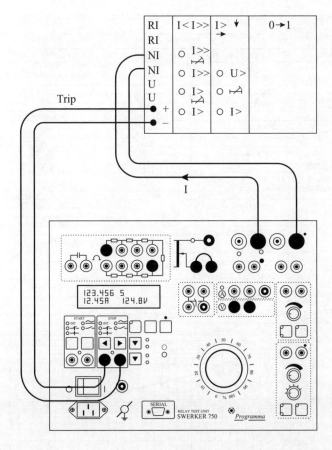

Figure 14.2   *Application example of a relay testing unit (Programma Sverker 750). (Reproduced by permission of GE Power Systems-Programma Electric AB)*

- operating times for different multipliers, as measured by calibration tests. These should be checked against the data provided by the manufacturer;
- test data for the instantaneous units;
- finally, the equipment used in the test should be recorded together with any relevant observations, plus details of the personnel who participated in the test.

It is important to note that the tests referred to beforehand correspond to steady-state conditions, and the equipment to carry them out is rather conventional. As a consequence of technological developments, more sophisticated equipment is now available to undertake tests using signals very similar to those that exist during fault conditions. Since relays are required to respond to the transient conditions of disturbed power systems, their real response can be assessed by simulating the signals that go to the relays under such conditions. Several manufacturers offer equipment to carry out dynamic-state and transient simulation tests. A dynamic-state test is one

GERS Consulting Engineers	TEST REPORT	DATE
		TESTED BY
PROJECT	50/51-50/51N PHASE AND GROUND OVERCURRENT RELAY	APPROVED BY

MANUFACTURER SEL	TYPE 551	LOCATION ER-1	SERIAL NUMBER 2000263013	CIRCUIT R-MSW1-M-B

1. SETTINGS

Parameter	Phase	Neutral/Ground
Current Transformer (CT prim. Amps)	600	600
Current Transformer (CT sec. Amps)	5	5
51 Relay Curve	U4 (Extremely Inverse)	U3 (Very Inverse)
51 Primary Pickup (Amps)	520	535
51 Tap	4.3	4.5
51 Time Dial	2.5	2
50 Primary Pickup (Amps)	Not Used	Not Used
50 Tap	Not Used	Not Used
50 Time Delay (cy)	Not Used	Not Used

2. PHASE UNIT TESTS

2.1. Overcurrent Pick up

Parameter	Phase	Theoretical	Result	% Error
	A		4.30	0.00%
Pick up current [A]	B	4.30	NA	
	C		4.31	0.23%

2.2. Check of Time Operation Curve

Parameter		Multiples of Pick Up Current					
		2.0	2.5	3.0	3.5	4.0	4.5
Injected current [A]	Fixed	8.60	10.75	12.90	15.05	17.20	19.35
Operation Time [s]	Theoretical	4.813	2.788	1.860	1.348	1.033	0.824
Operation Time [s] Phase A	Measured	4.90	2.75				
% Error-Phase A		1.81%	1.36%				
Operation Time [s] Phase B	Measured	NA	NA				
% Error-Phase B		NA	NA				
Operation Time [s] Phase C	Measured	4.79	2.78				
% Error-Phase C		0.48%	0.29%				

2.3. Operation Curve (Extremely Inverse)

2.4. Instantaneous Pick Up

Parameter	Phase	Theoretical	Result	% Error
	A			
Instantaneous Pick Up [A] Aprox	B	NA		
	C			

2.5. Instantaneous Pick Up

Parameter	Phase	Theoretical	Result	% Error
	A			
Operation Time [cycles]	B	NA		
	C			

2.6. Signaling Tests

Test	Phase	Signaling	Test	Phase	Signaling
	A	OK		A	NA
Time Overcurrent	B	OK	Instantaneous Overcurrent	B	NA
	C	OK		C	NA

3. TEST EQUIPMENT USED

4. REMARKS

	Initials/Signature	
	Reviewed	Approved

Figure 14.3   Typical test report sheet for overcurrent relays

in which phasor test quantities representing multiple power system conditions are synchronously switched between states. Power system characteristics, such as high frequency and DC decrement, are not, however, represented in this test. A transient simulation test signal can represent in frequency content, magnitude, and duration, the actual input signals received by a relay during power system disturbances.

Current transformer tests

Before commissioning a protection scheme, it is recommended that the following features of current transformers are tested.

Overlap of CTs

When CTs are connected in order that a fault on a breaker is covered by both protection zones, the overlap connections should be carefully checked. This should be carried out by a visual inspection. If this is not possible, or difficult, a continuity test between the appropriate relay and the secondary terminals of the appropriate CT should be carried out.

Correct connection of CTs

There are often several combinations of CTs in the same bushing and it is important to be sure that the CTs are correctly connected to their respective protection. Sometimes all the CTs have the same ratio but much different characteristics, or the ratios are different but the CTs are located close together, which can cause confusion.

Polarity

Each CT should be tested individually in order to verify that the polarity marked on the primary and secondary windings is correct. The measuring instrument connected to the secondary of the CT should be a high impedance voltmeter or a moving coil ammeter, with centre zero. A low voltage battery is used in series with a push-button switch in order to energise the primary. When the breaker is closed the measuring instrument should make a small positive deflection, and on opening the breaker there should be a negative deflection, if the polarity is correct.

Primary injection test

This test checks out all the protection system, including the current transformers. The principal aims of this test are to verify the CT transformation ratios and all the secondary circuit wiring of both the protection and the measurement CTs so that the operation of the tripping, signalling and alarm circuits is confirmed.

Figure 14.4 shows the schematic diagram of a typical equipment used for carrying out primary injection tests. The test current is usually between 100 and 400 A. The two high current terminals of the primary injection equipment marked <①> and <②> should be temporarily connected to the terminals of the CT that is being tested.

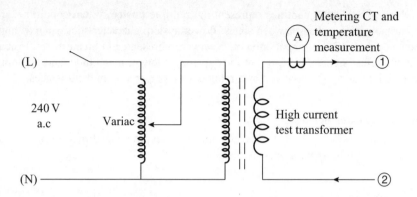

Figure 14.4   Equipment arrangement for primary injection test

14.2.3   *Periodic maintenance*

Protection equipment can be inactive for months; however, it is required to operate swiftly and accurately if a fault occurs on its associated primary equipment. Perhaps one of the most difficult requirements of the protection is that it should remain inoperative for external and close faults. Periodic maintenance on protection equipment should be carried out in order to be sure that the protection is always able to function correctly as and when required.

When a maintenance programme has been produced, the frequency of maintenance will depend upon the history of each item of equipment and any tendency to faults. Excessive testing should be avoided so that the maintenance programme will show up faults, not cause them. The frequency of maintenance, however, will vary greatly with the type of equipment. Certain items should be checked continuously, while others should be tested weekly, monthly and even once a year or more.

The classes of equipment that should be maintained are:

- protection relays;
- auxiliary control relays;
- annunciators and alarm systems;
- additional control devices (knobs, keys, interlocks, etc.);
- fault recorders if they are stand-alone devices.

It should be noted that, for an electrical utility where rationing energy could have grave consequences, taking equipment out of service for maintenance should be minimised by suitably programming the work in such a way that supplies are not jeopardised. Wherever possible, maintenance of protection equipment should be arranged to take place at the same time as the primary equipment is out of service.

Methodology

In order to achieve the maintenance objectives, a co-ordinated programme of work, evaluation and re-scheduling is required, as indicated below:

- set up a maintenance programme on a six month or one year basis, with tentative dates for carrying out work on individual items of equipment;
- evaluate the performance of the programme at the end of each six month period to obtain results for analysis;
- take into account the results of the previous period when preparing the new programme, and in particular the maintenance frequency for each item of equipment;
- re-schedule the preventative maintenance programme when immediate and corrective maintenance actions are required;
- co-ordinate with other maintenance groups beforehand, in the case of other equipment that may also require maintenance;
- file and analyse the results obtained with the aim of improving the performance indices, and reducing the costs.

Maintenance criteria

Within the maintenance programme, various categories of work can be grouped in general terms such as the following. *Preventative maintenance* is those jobs that are carried out on the basis of the maintenance programme. *Corrective maintenance* covers those jobs that are not programmed and require immediate attention.

Additionally, there are two different ways of achieving the maintenance work. *On line:* when, in order to carry out the work, it is not necessary to take out of service the high voltage equipment associated with the relays or instruments that need checking. *Off line:* when it is necessary to take the associated high voltage equipment out of service.

The principal sources of establishing the maintenance criteria are:

- recommendations made by the manufacturers;
- analysis of fault statistics;
- network voltage level;
- types of relay or instrument used;
- relative importance of the protected equipment;
- maintenance experience;
- previous behaviour of the equipment;
- type of fault, if a fault recorder operates.

Starting from the fact that the operating time of relays is small in comparison to the operating time of the power equipment, the fundamental principle is to carry out the least testing to give the best operation. This philosophy tends towards extending maintenance intervals, depending on the circumstances. The inspection interval depends on various factors, and principally on the history of the installation under test, the type of electrical protection used (electromagnetic, solid state, microprocessor or numerical), the importance of the equipment in relation to the power system and the consequences of an incorrect relay operation, and environmental conditions. On the basis of these factors, criteria for the frequency of maintenance in protection and control can be established, depending on whether this is preventative or corrective work.

14.3   Commissioning numerical protection

As indicated in Section 3.4, numerical protection is based on a number of microprocessors, which can be programmed to reproduce virtually any type of protection and relay. It represents important advantages in reliability, space, and economics compared to conventional protection arrangements. The method of calculating settings on numerical protection does not differ greatly from that used for conventional protection, which has been described in earlier chapters. However, there are big differences in the procedures to be followed during testing and commissioning this type of protection. When numerical protection is used, most of the wiring of a traditional relay arrangement is substituted by the programmable logic, with the great advantage of having multiple configurations and settings available as the numerical protection permits it. Consequently, programmable logic should be included along with all settings in the documentation for construction, commissioning, and testing purposes.

14.3.1   Setting the parameters

The method of setting numerical protection differs considerably from that used for electromechanical relays. Setting the parameters of a numerical protection, also known as parameterisation, involves selecting the functions of system configuration, protection, control, communication, alarms, and reporting features. As with conventional relays, settings for non-unit protections are determined on the basis of short-circuit and co-ordination studies. For unit protection, settings do not depend on co-ordination studies; therefore, machine constants, system parameters, and operation criteria are required in order to define the appropriate settings. Parameter settings are achieved by carrying out the following operations:

- selecting the required power system configuration and parameters of general operation;
- selecting the required protection functions;
- determining the setting groups;
- determining the dynamic or adaptive schemes;
- determining the relay logic and number of digital inputs and outputs;
- determining the configuration of reporting features and alarms.

Figure 14.5 represents the set-up system summary in a numerical protection.

The capability of manually or automatically changing the setting groups to meet the needs of the system is one of the great advantages of using numerical protection. A setting group may be changed immediately when the system topology changes with no compromise on the reliability of the system.

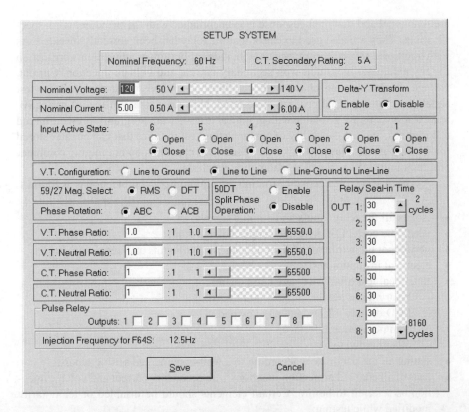

Figure 14.5   Setup system configuration of a numerical protection system (Beckwith M-3425). (Reproduced by permission of Beckwith Electric Co.)

14.3.2   Performance tests

Numerical protection devices are usually not allowed to be field calibrated since they are calibrated by specialised and certified process calibration equipment in the manufacturer's factory. Protective relay manufacturers perform numerous tests to each device before they are dispatched to customers after they pass the tests.

Most numerical relays, if not all, can be accessed through the front panel (Human Machine Interface – HMI) or through a serial port using a personal computer (PC). By accessing the relay with a PC it is possible to easily perform steady-state and dynamic tests. In some cases it is the only method to program logic schemes.

During the commissioning of numerical relays the following primary test methods should be carried out in order to check relay performance and operation values within the manufacturer's specifications.

(i)   Acceptance tests

The traditional acceptance tests are used to determine that the relay's system configuration, protection, control, metering, communication, alarms, and reporting features are functional and that their response is in accordance with the manufacturer's specifications. They are also used to check that the relay has been correctly installed. As with traditional relays, acceptance tests include calibration or adjustment verification for any parameter measured with secondary current and voltage injection. Current, voltage, frequency, and phase angle relationship applied to the device under test should be controlled and properly monitored. Nowadays, state-of-the-art protective relay test systems are available with graphical test software programs, pre-configured test routines, and the capability of personalising test routines that can be later stored for testing similar relays. Most of these software programs do not require computer programming skills to develop automatic relay tests.

A typical scope of acceptance routine tests for numerical protection is as follows:

- isolate all the relay inputs and outputs signals;
- verify the correct polarity and supply voltage value;
- verify that the wiring of all analogue and digital inputs and outputs is correct;
- electrical insulation test;
- functional test of all hardware components:
  - binary input/output;
  - LCD display, LEDs, and any key pad;
  - communication interface;
  - other hardware.
- setting all the parameters according to a well-documented setting report;
- tests of all enabled protection functions with secondary current and voltage injection as referred to before. Certain elements should be disabled or temporarily modified while testing;
- verification of operating measured-values;
- tests of the programmable logic scheme(s);
- test of targets and output contacts;
- verification of reporting features and alarms;
- reload/verify the in-service settings into the relay.

(ii)   Functional tests

After the acceptance tests have been performed and it has been verified that the relay can meet the intended specifications, functional tests should be performed. The specific procedure for each relay depends on the type of protection involved and the logic scheme implemented. A general procedure for carrying out these tests for a numerical protection arrangement is given below:

- verify external AC and DC input signals;
- verify external input contacts;
- verify tripping and signalling;
- verify remote/transferred tripping;
- verify interaction with the SCADA system, if applicable;
- verify setting group change according to the logic setting.

Functional tests can be easily achieved if the advanced tools for metering and reporting of the numerical protection are used. Some of these features are logic status report, in-service readings, metering data, event report, and oscillographies. The two last features are useful not only for commissioning, but also for troubleshooting.

(iii)   Dynamic tests

In a dynamic test the value of the applied phasors that represent the power system conditions should be properly adjusted when analysing pre-fault load, fault and post-fault states. In this test there are no typical characteristics of a power system, such as high frequency and DC decrement. Changes in phasor values (i.e. the magnitude and/or angle) should be controlled in such manner that they do not present any significant slip during the whole test.

(iv)   Transient simulation tests

The test signal of the transient simulation represents, for the relay, the actual external signals of frequency, magnitude, and duration when a disturbance occurs in the power system. These signals may involve offset (DC displacement), the saturation effects of CTs and the response of transient voltage surge suppressors.

(v)   End-to-end tests

End-to-end testing methodology is appropriate to verify the communication scheme and the protection system in an electrical line at the same time. Standard end-to-end testing methods use the following resources on each side of the line:

- three-phase protection relay test equipment with appropriate communication and transient simulation capabilities;
- for the time synchronisation of the two, test equipment used could be one of the following methods: GPS-satellite receiver (currently the most common method), pilot wire, fibre optic, or with the power system;
- fault simulation such as the Electromagnetic Transients Program (EMTP) or real transient files taken from disturbance recorders.

The Electromagnetic Transients Program is a computer program for simulating high speed transient effects in electric power systems. It includes a wide range of modelling capabilities including oscillations ranging in duration from microseconds to seconds.

Disturbance recorders generate oscillographic records in IEEE Standard COMTRADE – Common Format for Transient Data Exchange – files. These files permit a comprehensive analysis of a particular event to be achieved by using appropriate software and are fed into relay test equipment which can reproduce the original waveform to be injected into a relay under test.

Bibliography

ANDERSON, P. M.: 'Analysis of faulted power systems' (The Iowa State University Press, Ames, Iowa, USA, 1995)

ANDERSON, P. M., and MIRHEYDAR, M.: 'An adaptive method of setting under-frequency load shedding relays', *IEEE Trans.*, 1992, **PWRS-7**(2), pp. 647–655

ANSI/IEE standard C57.13: 'Standard requirements for instrument transformers', 1978

ANSI/IEEE standard 141: 'Recommended practice for electric power distribution for industrial plants', 1990

ANSI/IEEE standard 242: 'Recommended practice for protection and co-ordination of industrial and commercial power systems', 1986

ANSI/IEEE standard 399: 'Recommended practice for power system analysis', 1990

BLACKBURN, J. L.: 'Protective relaying' (Marcel Dekker, New York, 1998, 2nd edn.)

BURNETT, J.: 'IDMT relay tripping of main incoming circuit breakers', *Power Eng., J.*, 1990, **4**, pp. 51–56

CLOSSON, J., and YOUNG, M.: 'Commissioning numerical relays', XIV IEEE Summer Meeting, Acapulco, Mexico, July 2001

COOK, V.: 'Analysis of distance protection' (John Wiley & Sons, New York, 1985)

DAVIES, T.: 'Protection of industrial power systems' (Pergamon Press, Cleveland, USA, 1983)

'Developments in power system protection, 6th international conference', *IEE Conf. Publ.*, **434**, 1997

'Developments in power system protection, 7th international conference', *IEE Conf. Publ.*, **479**, 2001

'Distribution protection', ABB buyer guide, vols. I, II and III, 1989–1990

'Electrical distribution system protection manual' (Cooper Power Systems, USA, 1990, 3rd edn.)

ELMORE, W. A.: 'Protective relaying theory and applications' (ABB, New York, 1994)

FEENAN, J.: 'The versatility of high-voltage fuses in the protection of distribution systems', *Power Eng. J.*, 1987, **1**, pp. 109–115

GERS, J. M.: 'Aplicación de protecciones eléctricas a sistemas de potencia' ('Application of electrical protection to power systems'), (Universidad de Valle, Cali, 1993)

GERS, J. M.: 'Enhancing numerical relaying performance with logic customization', XV IEEE Summer meeting, Acapulco, Mexico, July 2002

GROSS, C.A.: 'Power system analysis' (John Wiley & Sons, New York, 1986, 2nd edn.)

'Guides and standards for protective relaying systems' (IEEE, New York, 1995)

HARKER, K.: 'Power system commissioning and maintenance practice' (Peter Peregrinus, London, 1998)

HEADLEY, A., BURDIS, E. P., and KELSEY, T.: 'Application of protective devices to radial overhead line networks', *IEE Proc. C, Gener. Transm. Distrib.*, 1986, **133**, pp. 437–440

HORAK, J.: 'Pitfalls and benefits of commissioning numerical relays' (Neta World, Colorado, 2003)

'Instruction manual for generator protection system BE1-GPS100', Basler Electric, Publication 9 3187 00 990, Revision C, November 2001

'Instruction manual for overcurrent BEI-951', Basler Electric, Publication 9 3289 00 990, Revision F, August 2002

'Instruction Book M-3311 Transformer Protection Relay', Beckwith Electric Co., Publication 800-3311-IB-01MC1, February 2003

'Instruction Book M-3425 Generator Protection Relay', Beckwith Electric Co., Publication 800-3425-IB-02MC2, January 2003

LAKERVI, E., and HOLMES, E. J.: 'Electricity distribution network design' (Peter Peregrinus, 1995, 2nd edn.; revised 2003)

LAYCOCK, W. J.: 'Management of protection', *Power Eng. J.*, 1991, **5**, pp. 201–207

LO, K. L., McDONALD, J. R., and YOUNG, D. J.: 'Expert systems applied to alarm processing in distribution control centres', in Conference proceedings, UPEC 89, Belfast, 1989

LO, K. L., and NASHID, L.: 'Expert systems and their applications to power systems, Part 3 Examples of application', *Power Eng. J.*, 1993, **7**(5), pp. 209–213

LONG, W., COTCHER, D., RUIU, D., ADAM, P., LEE, S., and ADAPA, R.: 'EMTP – a powerful tool for analysing power system transients', *Computer applications in power*, IEEE, 1990, **3**(3), pp. 36–41

'Microprocessor relays and protection systems', IEEE Tutorial Course, 88EH0269-1-PWR, 1987

MOORE, P. J., and JOHNS, A. T.: 'Adaptive digital distance protection', *IEE Conf. Publ.*, 1989, **302**, pp. 187–191

PHADKE, A. G., and HOROWITZ, S. H.: 'Adaptive relaying', *IEEE Comput. Appl. Power*, 1990, **3**(3), pp. 47–51

PHADKE, A. G., and THORP, J. S.: 'Computer relaying for power systems' (John Wiley & Sons, New York, 1988)

'Power system protection' (Peter Peregrinus, 1997)

'Protection and control devices standards, dimensions and accessories', Basler Electric, Product Bulletin SDA-5 8-01, August 2001

'Protective relays application guide', GEC Measurements (Baldini and Mansell, 1987, 3rd edn.)

'Quadramho static distance protection relay', GEC Measurements, publication R-5580B, London

'Relay Testing Unit Sverker 750', Programma Electric ABB, User's Manual, Ref. No. ZP-CD01E R01A, 1994

SIDHU, T. S., SACHDEV, M. S., and WOOD, H. C.: 'Microprocessor-based relay for protecting power transformers', *IEE Proc. C, Gener. Transm. Distrib.*, 1990, **137**, pp. 436–444

STEVENSON, W. D.: 'Elements of power system analysis' (McGraw Hill, New York, 1982, 4th edn.)

'Switching, protection and distribution in low voltage networks', Siemens, (Publicis MCD Verlag, 1994, 2nd edn.)

TEO, C. Y., and CHAN, T. W.: 'Development of computer-aided assessment for distribution protection', *Power Eng. J.*, 1990, **4**, pp. 21–27

URDANETA, A. J., RESTREPO, H., MARQUEZ, J., and SANCHEZ, J.: 'Co-ordination of directional overcurrent relay timing using linear programming', IEEE PAS Winter meeting, New York, February 1995

WHITING, J. P., and LIDGATE, D.: 'Computer prediction of IDMT relay settings and performance for interconnected power systems', *IEE Proc. C., Gener. Transm. Distrib.*, 1983, **130**(3), pp. 139–147

WRIGHT, A., and NEWBERRY, P. G.: 'Electric fuses' (Peter Peregrinus, 1994, 2nd edn.)

WRIGHT, A.: 'Application of fuses to power networks', *Power Eng. J.*, 1991, **4**, pp. 293–296, ibid. 1991, **5**, pp. 129–134

YOUNG, D. J., LO, K. L., McDONALD, J. R., HOWARD, R., and RYE, J.: 'Development of a practical expert system for alarm processing', *IEE Proc. C, Gener. Transm. Distrib.*, 1992, **139**(5), pp. 437–447

ZHANG, Z. Z., and CHEN, D.: 'An adaptive approach in digital distance protection', *IEEE Trans.*, 1991, **PWRD-6**(1), pp. 135–142

Appendix

Solutions to exercises

To assist in the understanding of the solutions, the original diagrams included in the Exercises have been redrawn, and where necessary, additional diagrams are included to illustrate the calculations.

Solution to Exercise 1.1

Fault F_1 was correctly cleared by the tripping of breakers 2 and 5; therefore breakers 3 and 4 operated incorrectly and should be entered in column 3.

For fault F_2, breakers 25 and 26 should have opened to clear the fault. These are not shown in column 2; therefore they should be inserted in column 3 as having mal-operated. Breakers 21 and 22 should have been tripped by back-up protection and shown in column 5. Breakers 23 and 24 should not have tripped as the flow of fault current was against their directional settings and these two breakers should be placed in column 3 as having mal-operated. With the failure of breaker 25 to trip, the fault was finally cleared by breaker 27 tripping on back-up protection, which should be entered in column 5 also.

Fault F_3 should have been tripped by opening breakers 11 and 17. These are shown as having operated in column 2. Therefore, the tripping of breakers 10 and 19 was unnecessary, and these two breakers should be placed in column 3.

The completed table is shown in Table A.1.

Solution to Exercise 5.1

The equivalent impedances of all the circuits are calculated, referred to the same voltage level, to obtain the diagram of the positive-sequence network and thus calculate the short-circuit levels.

Table A.1   Relay/breaker operation for Exercise 1.1

Case	Breakers that operated	Breakers that mal-operated	Tripped by primary protection	Tripped by back-up protection
F_1	2, 3, 4, 5	3, 4	2, 5	–
F_2	21, 22, 23, 24, 27	25, 26, 23, 24	–	21, 22, 27
F_3	10, 11, 17, 19	10, 19	11, 17	–

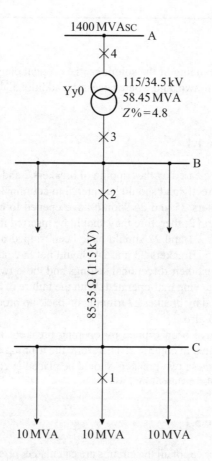

Figure A.1   Schematic diagram for Exercise 5.1

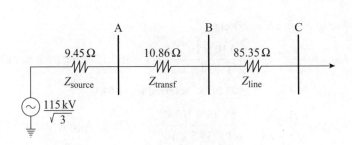

Figure A.2   Equivalent circuit of the system shown in Exercise 5.1

$$Z_{source} = \frac{V^2}{P_{sc}} = \frac{(115 \times 10^3)^2}{1400 \times 10^6} = 9.45\,\Omega,\ \text{referred to 115 kV}$$

$$Z_{transf} = Z_{pu(transf)} \times Z_{base(transf)} = 0.048 \times \frac{(115 \times 10^3)^2}{58.45 \times 10^6}$$

$$= 10.86\,\Omega,\ \text{referred to 115 kV}$$

$$Z_{line} = 85.35\,\Omega,\ \text{referred to 115 kV}$$

The equivalent circuit is shown in Figure A.2.

Calculation of short-circuit currents

Busbar A:

$$I_{sc(A)} = \frac{115 \times 10^3}{\sqrt{3} \times Z_{source}} = \frac{115 \times 10^3}{\sqrt{3} \times 9.45} = 7025.96\,\text{A},\ \text{referred to 115 kV}$$

Busbar B:

$$I_{sc(B)} = \frac{115 \times 10^3}{\sqrt{3} \times (Z_{source} + Z_{transf})} = \frac{115 \times 10^3}{\sqrt{3} \times (9.45 + 10.86)}$$

$$= 3269.09\,\text{A},\ \text{referred to 115 kV}$$

$$= 3269.09 \times \frac{115 \times 10^3}{34.5 \times 10^3} = 10896.97\,\text{A},\ \text{referred to 34.5 kV}$$

Busbar C:

$$I_{sc(C)} = \frac{115 \times 10^3}{\sqrt{3} \times (Z_{source} + Z_{transf} + Z_{line})}$$

$$= \frac{115 \times 10^3}{\sqrt{3} \times (9.45 + 10.86 + 85.35)} = 628.39\,\text{A},\ \text{referred to 115 kV}$$

$$= 628.39 \times \frac{115 \times 10^3}{34.5 \times 10^3} = 2094.62\,\text{A},\ \text{referred to 34.5 kV}$$

Calculation of nominal currents

Relay 1: $I_{nom1} = \dfrac{P_{nom1}}{\sqrt{3} \times V_1} = \dfrac{10 \times 10^6 \text{ VA}}{\sqrt{3} \times 34.5 \times 10^3 \text{ V}} = 167.35 \text{ A, referred to } 115 \text{ kV}$

Relay 2: $I_{nom2} = 3 \times I_{nom1} = 502.04 \text{ A, referred to } 34.5 \text{ kV}$

Relay 3: $I_{nom3} = \dfrac{P_{nom3}}{\sqrt{3} \times V_3} = \dfrac{58.45 \times 10^6 \text{ VA}}{\sqrt{3} \times 34.5 \times 10^3 \text{ V}} = 978.15 \text{ A, referred to } 34.5 \text{ kV}$

Relay 4: $I_{nom4} = \dfrac{P_{nom4}}{\sqrt{3} \times V_4} = \dfrac{58.45 \times 10^6 \text{ VA}}{\sqrt{3} \times 115 \times 10^3 \text{ V}} = 293.44 \text{ A, referred to } 115 \text{ kV}$

Selection of CT transformation ratios

The CT ratio is chosen from the larger of the two following values:

(i) the nominal current.
(ii) the maximum short-circuit current without saturation occurring ($0.05 \times I_{sc}$).

The values of the CT ratios calculated in accordance with the abovementioned criteria are given in Table A.2.

Setting of instantaneous units

The instantaneous units have settings from 6 to 144 A in steps of 1 A. The setting of the instantaneous units on the feeders are based on $0.5 \times I_{sc}$ at busbar C.

Relay 1:

$I_{inst} = 0.5 \times I_{sc(C)} = 0.5 \times 2094.62 = 1047.31 \text{ primary amps}$

$\quad\quad = 1047.31 \times (5/200) = 26.18 \text{ secondary amps.}$

Set at 27 secondary amps, equivalent to 1080 primary amps.

Table A.2   Summary of calculations for CT ratios for Exercise 5.1

Relay number	P_{nom} (MVA)	I_s (A)	$0.05 \times I_{sc}$ (A)	I_{nom} (A)	CT ratio
1	10.00	2094.62	104.78	167.35	200/5
2	30.00	10896.97	544.85	502.04	600/5
3	58.45	10896.97	544.85	978.15	1000/5
4	58.45	7025.96	351.30	293.44	400/5

Relay 2:
The setting is made by taking 125 per cent of the value of the current for the maximum fault level which exists at the following substation.

$$I_{inst} = 1.25 \times I_{sc(C)} = 1.25 \times 2094.62 = 2618.28 \text{ primary amps}$$

$$= 2618.28 \times (5/600) = 21.82 \text{ secondary amps.}$$

Set at 22 secondary amps, equivalent to 2640 primary amps.

Relay 3:
The instantaneous unit is overridden in order to avoid lack of co-ordination with the transformer.

Relay 4:
The setting is made by taking 125 per cent of the value of the short-circuit current existing on the 34.5 kV side, referred to the 115 kV side.

$$I_{inst} = 1.25 \times I_{sc(B)}, \text{ referred to } 115 \text{ kV}, = 1.25 \times 3269.09$$

$$= 4086.36 \text{ primary amps}$$

$$= 4086.36 \times (5/400) = 51.08 \text{ secondary amps.}$$

Set at 52 secondary amps, equivalent to 4160 primary amps.

Selection of pick-up settings

From eqn. (5.5), pick-up setting $(PU) = OLF \times I_{nom} \times (1/CTR)$.
Using an overload factor of 1.5, the pick-up settings are:
Relay 1: $PU_1 = 5 \times 167.35 \times (5/200) = 6.28$; set at 7.
Relay 2: $PU_2 = 1.5 \times 502.04 \times (5/600) = 6.28$; set at 7.
Relay 3: $PU_3 = 1.5 \times 978.15 \times (5/1000) = 7.34$; set at 8.
Relay 4: $PU_4 = 1.5 \times 293.44 \times (5/400) = 5.5$; set at 6.

Time dial settings

Relay 1:
The minimum time dial setting is chosen for this relay, since it is at the end of the circuit and therefore does not need to be co-ordinated with any other protective device. Use a time dial setting of 1/2.

The operating time of relay 1 should be calculated from just before the operation of its instantaneous unit. Based on the primary setting of 1080 A and a pick-up setting of 7, the plug setting multiplier (PSM) is

$$PSM = I_{inst.prim.1} \times (1/CTR_1) \times (1/PU_1) = 1080 \times (5/200) \times (1/7)$$

$$= 3.86 \text{ times.}$$

With a time dial setting of 1/2, and PSM = 3.86 times, from the relay curves, $t_1 = 0.16$ s.

Relay 2:
Calculate the back-up time to relay 1; $t_2 = 0.16 + 0.4 = 0.56$ s.

Calculate the multiplier on the basis of the 1080 primary amps for the CT associated with relay 1.

$$PSM_{2a} = 1080 \times (1/CTR_2) \times (1/PU_2) = 1080 \times (5/600) \times (1/7) = 1.29 \text{ times.}$$

With $PSM_{2a} = 1.29$ times, and $t_{2a} = 0.56$ s, from the relay curves, the time dial setting $= 1/2$.
The operating time of relay 2 should be calculated from just before the operation of its instantaneous unit:

$$PSM_{2b} = I_{\text{inst.prim.2}} \times (1/CTR_2) \times (1/PU_2) = 2640 \times (5/600) \times (1/7)$$
$$= 3.14 \text{ times.}$$

With a time dial setting of 1/2, and $PSM_{2b} = 3.14$ times, from the relay curves, $t_{2b} = 0.25$ s.

Relay 3:
Calculate the back-up time to relay 2; $t_{3a} = 0.25 + 0.4 = 0.65$ s.
Calculate the multiplier on the basis of the 2640 primary amps for the CT associated with relay 2.

$$PSM_{3a} = 2640 \times (1/CTR_3) \times (1/PU_3) = 2640 \times (5/1000) \times (1/8)$$
$$= 1.65 \text{ times.}$$

With $PSM_{3a} = 1.65$ times, and $t_{3a} = 0.65$ s, from the relay curves, time dial setting $= 1/2$.
As the instantaneous unit is overridden the multiplier is obtained using the short-circuit current:

$$PSM_{3b} = I_{\text{sc.3}} \times (1/CTR_3) \times (1/TAP_3) = 10896.97 \times (5/1000) \times (1/8)$$
$$= 6.81 \text{ times.}$$

With a time dial setting of 1/2, and $PSM = 6.81$ times, from the relay curves, $t_3 = 0.07$ seconds.

Relay 4:
Calculate the back-up time to relay 3; $t_4 = 0.07 + 0.4 = 0.47$ s.
Calculate the multiplier on the basis of the 10896.97 primary amps for the CT associated with relay 3, but referred to 115 kV:

$$PSM_4 = 10896.97 \times (34.5/115) \times (1/CTR_4) \times (1/PU_4)$$
$$= 10806.97 \times (34.5/115) \times (5/400) \times (1/6) = 6.81 \text{ times.}$$

With $PSM_4 = 6.81$ times, and $t_4 = 0.47$ s, from the relay curves, time dial setting $= 4$.
The settings are summarised in Table A.3.

Table A.3   Summary of relay settings for Exercise 5.1

Relay number	CT ratio	Pick-up (A)	Time dial	Instantaneous units	
				primary (A)	secondary (A)
1	200/5	7	1/2	1080	27
2	600/5	7	1/2	2640	overridden
3	1000/5	8	1/2	overridden	overridden
4	400/5	8	4	4160	52

Solution to Exercise 5.2

Figure A.3 gives the single line diagram of the substation.

Calculation of short-circuit currents

13.2 kV busbar:
From Figure A.3, which repeats Figure 5.28, it can be noted that the short-circuit level at the 13.2 kV busbar is 21900 amperes, and from this the equivalent impedance of the system behind the busbar can be obtained:

$$I_{sc(13.2\,kV)} = 21900\,A, \text{ referred to } 13.2\,kV$$

$$P_{sc} = \sqrt{3} \times 13200\,V \times 21900\,A = 500.70125\,MVA$$

$$Z_{source} = \frac{V^2}{P_{sc}} = \frac{(13.2 \times 10^3)^2}{500.70125 \times 10^6} = 0.348\,\Omega \text{ referred to } 13.2\,kV$$

$$= 0.348 \times \left(\frac{115 \times 10^3}{13.2 \times 10^3}\right)^2 = 26.414\,\Omega \text{ referred to } 115\,kV$$

115 kV busbar:
In order to obtain this short-circuit fault level it is necessary to know the equivalent impedance behind this busbar:

$$Z'_{source} + Z_{transf} = Z_{source}$$

where Z'_{source} is the equivalent impedance behind the 115 kV busbar and Z_{transf} is the equivalent impedance of the two transformers.

$$Z_{transf1} = 0.0963 \times \frac{(115 \times 10^3)^2}{20 \times 10^6} = 63.6784\,\Omega \text{ referred to } 115\,kV$$

$$Z_{transf3} = 0.101 \times \frac{(115 \times 10^3)^2}{41.7 \times 10^6} = 32.0318\,\Omega \text{ referred to } 115\,kV$$

The equivalent impedance of the two transformers is $Z_{transf} = 63.6784||32.0318 = 21.3116\,\Omega$ referred to 115 kV. Thus, $Z'_{source} + Z_{transf} = 26.414\,\Omega$ referred to 115 kV. $Z'_{source} = 26.414 - 21.3116\,\Omega, = 5.1024\,\Omega$ referred to 115 kV.

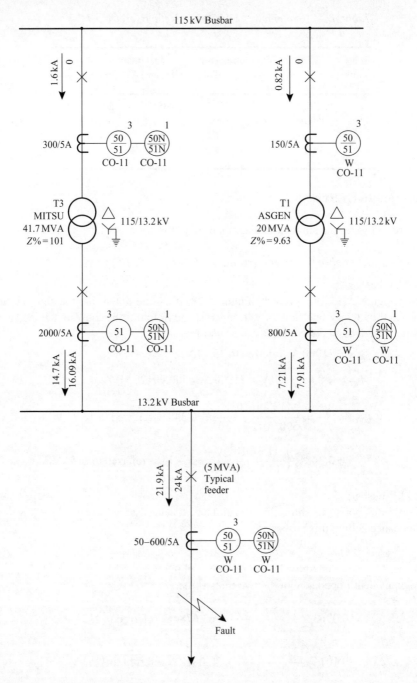

Figure A.3   Single-line diagram of substation for Exercise 5.2

With this value of equivalent impedance the short-circuit level at the 115 kV busbar can be obtained:

$$I_{sc(115\,kV)} = \frac{115 \times 10^3}{\sqrt{3} \times 5.1024} = 13012.56\,A \text{ referred to } 115\,kV$$

Calculation of nominal currents

Relay on the high voltage side of transformer T1:

$$I_{nom(HV.T1)} = \frac{P_{nom.trans1}}{\sqrt{3} \times V} = \frac{20 \times 10^6\,VA}{\sqrt{3} \times 115 \times 10^3\,V} = 100.41\,A \text{ referred to } 115\,kV$$

Relay on the low voltage side of transformer T1:

$$I_{nom(LV.T1)} = \frac{P_{nom.trans1}}{\sqrt{3} \times V} = \frac{20 \times 10^6\,VA}{\sqrt{3} \times 13.2 \times 10^3\,V} = 874.77\,A \text{ referred to } 13.2\,kV$$

Relay on the high voltage side of transformer T3:

$$I_{nom(HV.T3)} = \frac{P_{nom.trans3}}{\sqrt{3} \times V} = \frac{41.7 \times 10^6\,VA}{\sqrt{3} \times 115 \times 10^3\,V} = 209.35\,A \text{ referred to } 115\,kV$$

Relay on the low voltage side of transformer T3:

$$I_{nom(LV.T3)} = \frac{P_{nom.trans3}}{\sqrt{3} \times V} = \frac{41.7 \times 10^6\,VA}{\sqrt{3} \times 13.2 \times 10^3\,V} = 1823.9\,A \text{ referred to } 13.2\,kV$$

Relay on a typical feeder:

$$I_{nom(feeder)} = \frac{P_{nom.feeder}}{\sqrt{3} \times V} = \frac{5 \times 10^6\,VA}{\sqrt{3} \times 13.2 \times 10^3\,V} = 218.69\,A \text{ referred to } 13.2\,kV$$

Setting of instantaneous units

Relay on typical feeder:
The setting of the instantaneous units on relays associated with feeders are calculated using the mid value of the short circuit current seen by the relay.

$$I_{inst} = 0.5 \times I_{sc(13.2\,kV\,bus)} = 0.5 \times 21900 \text{ primary amps} = 10950\,A$$

Relay on the high voltage side of transformer T1:
$I_{inst} = 1.25 \times I_{sc(13.2\,kV\,bus)}$, referred to 115 kV, $= 1.25 \times 7210 \times (13.2/115) = 1034.48$ primary amps, $\Rightarrow 1034.48 \times (5/150) = 34.8$ secondary amps. Set relay to 35 secondary amps, equivalent to 1050 primary amps.

Relay on the low voltage side of transformer T1:
The instantaneous element is overridden to avoid the possibility of loss of discrimination with the transformer secondary protection.

Relay on the high voltage side of transformer T3:
The setting is based on 125 per cent of the short circuit current that exists at the 13.2 kV busbar seen by the relay in question, referred to the 115 kV side.

$I_{\text{inst}} = 1.25 \times I_{\text{sc}(13.2\text{kV})}$, referred to $115\,\text{kV}, = 1.25 \times 14700 \times (13.2/115) = 2109.13$ primary amps, $\Rightarrow 2109.13 \times 5/300 = 35.15$ secondary amps. Set at 36 secondary amps, equivalent to 2160 primary amps.

Relay on the low voltage side of transformer T3:
The instantaneous element is overridden to avoid the possibility of loss of discrimination with the transformer secondary protection.

Selection of pick-up settings

Relay on typical feeder:

$\text{PU}_{\text{feeder}} = 1.5 \times 218.69 \times (5/300) = 5.47$; $\text{PU}_{\text{feeder}}$ set at 6.

Relay on high voltage side of transformer T1:

$\text{PU}_{\text{HV.T1}} = 1.5 \times 100.41 \times (5/150) = 5.02$; $\text{PU}_{\text{HV.T1}}$ set at 6.

Relay on low voltage side of transformer T1:

$\text{PU}_{\text{LV.T1}} = 1.5 \times 874.77 \times (5/900) = 7.28$; $\text{PU}_{\text{LV.T1}}$ set at 8.

Relay on high voltage side of transformer T3:

$\text{PU}_{\text{HV.T3}} = 1.5 \times 209.35 \times (5/200) = 5.23$; $\text{PU}_{\text{HV.T3}}$ set at 6.

Relay on low voltage side of transformer T3:

$\text{PU}_{\text{LV.T3}} = 1.5 \times 1823.9 \times (5/2000) = 6.84$; $\text{PU}_{\text{LV.T3}}$ set at 7.

Time dial settings

Relay on typical feeder:
Use the minimum time dial setting, in order to deal with relay at the end of the circuit; time dial setting $= 1/2$.
It is necessary to calculate the relay operating time, just before the instantaneous unit operates:

$$\text{PSM}_{\text{feeder}} = I_{\text{inst.prim}} \times (1/\text{CTR}) \times (1/\text{PU}_{\text{feeder}}) = 10950 \times (5/300) \times (1/6)$$

$$= 30 \text{ times.}$$

With time dial setting of 1/2, and $\text{PSM}_{\text{feeder}} = 30$, from the relay operating curve $t_{\text{feeder}} = 0.024\,\text{s}$.

Relay on low voltage side of transformer T1:
The back-up time over the feeder relay is now calculated; $t_{\text{LV.T1}} = 0.024 + 0.4 = 0.424\,\text{s}$. The multiplier for this relay is based on the 10950 primary amps for the CT associated with the feeder:

$$\text{PSM}_{\text{LV.T1}} = 10950 \times (1/\text{CTR}_{\text{LV.T1}}) \times (1/\text{PU}_{\text{LV.T1}}) = 10950 \times (5/900) \times (1/8)$$

$$= 7.6 \text{ times.}$$

With $PSM_{LV.T1} = 7.6$ and the relay back-up time $= 0.424$ s, from the relay curves, time dial setting $= 4$.

As the instantaneous unit is overridden, the short-circuit current seen by the relay and multiplied by 0.86 is used for the delta-star transformer.

$$PSM = 0.86 \times I_{sc.LV.T1} \times (1/CTR_{LV.T1}) \times (1/PU_{LV.T1})$$

$$= 0.86 \times 7210 \times (5/900) \times (1/8) = 4.3 \text{ times.}$$

With a time dial setting of 4, and $PSM = 4.3$, from the relay curve, $t_{HV.T1} = 1.4$ s.

Relay on the high voltage side of transformer T1:
Required operating time, as back up to the low voltage side relay, $= 1.4 + 0.4 = 1.8$ s.
The multiplier is calculated on the basis of the 7210 primary amps (referred to 115 kV) which is seen by the CT associated with the relay on the low voltage side of T1:

$$PSM_{HV.T1} = 7210 \times (13.2 \text{ kV}/115 \text{ kV}) \times (1/CTR_{HV.T1}) \times (1/PU_{HV.T1})$$

$$= 7210 \times (13.2 \text{ kV}/115 \text{ kV}) \times (5/150) \times (1/6) = 4.6 \text{ times}$$

With $PSM_{HV.T1} = 4.6$, and the back-up time $= 1.8$ s, from the relay curve, time dial setting $= 6$.

Relay on low voltage side of transformer T3:
The back-up time over the feeder relay is $0.024 + 0.4 = 0.424$ s.
The multiplier is calculated on the basis of the 10950 primary amps which is seen by the CT associated with the relay on the low voltage side of T1:

$$PSM_{LV.T3} = 10950 \times (1/CTR_{LV.T3}) \times (1/PU_{LV.T3}) = 10950 \times (5/2000) \times (1/7)$$

$$= 3.9 \text{ times.}$$

With $PSM_{LV.T3} = 3.9$, and the back-up time $= 0.424$ s, from the relay curve time dial setting $= 2$.

As the instantaneous unit is overridden, the short-circuit current seen by the relay on the low voltage side of T3 and multiplied by 0.86 is used for the delta-star transformer.

$$PSM = 0.86 \times I_{sc.LV.T3} \times (1/CTR_{LV.T3}) \times (1/PU_{LV.T3})$$

$$= 0.86 \times 14700 \times (5/2000) \times (1/7) = 4.5 \text{ times.}$$

With a time dial setting of 2, and $PSM = 4.5$, from the relay curve, $t_{LV.T3} = 0.6$ s.

Relay on the high voltage side of transformer T3:
Required operating time, as back up to the low voltage side relay, $= 0.6 + 0.4 = 1.0$ s.
The multiplier is calculated on the basis of the 14700 amps (referred to 115 kV) which is seen by the CT associated with the relay on the low voltage side of T3:

$$PSM_{HV.T3} = 14700 \times (13.2 \text{ kV}/115 \text{ kV}) \times (1/CTR_{HV.T3}) \times (1/PU_{HV.T3})$$

$$= 14700 \times (13.2 \text{ kV}/115 \text{ kV}) \times (5/300) \times (1/6) = 4.7 \text{ times}$$

With $PSM_{HV.T3} = 4.7$, and the back-up time $= 1.0$ s, from the relay curve, time dial setting $= 4$.

Solution to Exercise 5.3

The single diagram for the network being studied is shown in Figure A.4.

Maximum short-circuit currents

The equivalent impedances of all the circuits are calculated, referred to the same voltage level, in order to produce the diagram of positive-sequence impedances for calculating the short-circuit levels.

$$Z_{\text{source}} = \frac{V^2}{P_{\text{sc}}} = \frac{(115000)^2}{2570.87} = 5.144\,\Omega \text{ referred to } 115\,\text{kV},$$

$$= 5.144 \times \left(\frac{34.5}{115}\right)^2 = 0.463\,\Omega \text{ referred to } 34.5\,\text{kV}$$

$$Z_{\text{transf1}} = 0.117 \times \frac{(34500)^2\ \text{V}}{10.5 \times 10^6\ \text{VA}} = 13.26\,\Omega \text{ referred to } 34.5\,\text{kV}$$

$$Z_{\text{transf2}} = 0.06 \times \frac{(34500)^2\ \text{V}}{5.25 \times 10^6\ \text{VA}} = 13.6\,\Omega \text{ referred to } 34.5\,\text{kV}$$

$$Z_{\text{lineBC}} = 0.625\,\Omega/\text{km} \times 14.2\,\text{km} = 8.875\,\Omega \text{ at } 34.5\,\text{kV}$$

The equivalent circuit is shown in Figure A.5.

Calculation of short-circuit currents

Busbar A:

$$I_{\text{sc(A)}} = \frac{34.5 \times 10^3}{\sqrt{3}\,(0.463 + 13.26 + 8.875 + 13.6)} = 881.43\,\text{A referred to } 34.5\,\text{kV}$$

$$= 550.26 \times (34.5/13.2) = 1438.18\,\text{A referred to } 13.2\,\text{kV}$$

Busbar B:

$$I_{\text{sc(B)}} = \frac{34.5 \times 10^3}{\sqrt{3}\,(0.463 + 13.26 + 8.875)} = 881.43\,\text{A referred to } 34.5\,\text{kV}$$

Busbar C:

$$I_{\text{sc(C)}} = \frac{34.5 \times 10^3}{\sqrt{3}\,(0.463 + 13.26)} = 1451.47\,\text{A referred to } 34.5\,\text{kV}$$

Busbar D:

$$I_{\text{sc(D)}} = 12906.89\,\text{A referred to } 115\,\text{kV}$$

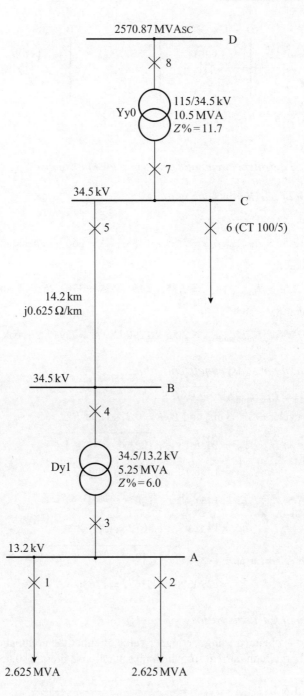

Figure A.4   Single-line diagram for Exercise 5.3

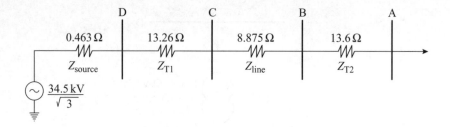

Figure A.5   Equivalent circuit of the system shown in Exercise 5.3

Calculation of maximum peak values

Circuit breaker 1:

$$I_{peak} = 2.55 \times I_{rms.sym.bkr1} = 2.55 \times 1438.14\,\text{A} = 3667.36 \text{ peak amperes}$$

Circuit breaker 5:

$$I_{peak} = 2.55 \times I_{rms.sym.bkr5} = 2.55 \times 1451.47\,\text{A} = 3701.25 \text{ peak amperes}$$

Circuit breaker 8:

$$I_{peak} = 2.55 \times I_{rms.sym.bkr8} = 2.55 \times 12906.89\,\text{A} = 32912.57 \text{ peak amperes}$$

Calculation of r.m.s. asymmetrical values

$$I_{rms.asym} = I_{rms.sym.int}\sqrt{2e^{-2(R/L)t} + 1}$$

where $t = 5$ cycles $= 83.33$ ms, and $L/R = 0.2$, so that $R/L = 5$.

$$\text{Breaker 1: } I_{rms.asym.} = 1438.18 \times \sqrt{\left(2e^{-2\times0.08333\times5}\right) + 1}$$

$$= 1438.18 \times 1.3672 = 1966.27\,\text{A}$$

$$\text{Breaker 5: } I_{rms.asym.} = 1451.47 \times \sqrt{\left(2e^{-2\times0.08333\times5}\right) + 1}$$

$$= 1451.47 \times 1.3672 = 1984.45\,\text{A}$$

$$\text{Breaker 8: } I_{rms.asym.} = 12906.89 \times \sqrt{\left(2e^{-2\times0.08333\times5}\right) + 1}$$

$$= 12906.89 \times 1.3672 = 17646.3\,\text{A}$$

Calculation of CT turns ratios

The CT ratio is obtained using the higher value of either the nominal current, or the maximum short circuit current for which no saturation is present ($0.05 \times I_{sc}$).

$$\text{Relays 1 \& 2: } I_{nom1} = I_{nom2} = \frac{P_{nom1}}{\sqrt{3}V_1} = \frac{2.625 \times 10^6\,\text{VA}}{\sqrt{3} \times 13.2 \times 10^3\,\text{V}}$$

$$= 114.81\,\text{A referred to } 13.2\,\text{kV}$$

Table A.4   Summary of CT ratio calculations for Exercise 5.3

Relay number	P_{nom} (MVA)	I_s (A)	$0.05 \times I_{sc}$ (A)	I_{nom} (A)	CT ratio
1 and 2	2.625	1438.18	71.91	114.81	150/5
3	5.25	1438.18	71.91	229.63	250/5
4	5.25	881.43	44.07	87.86	100/5
5	5.25	1451.47	72.57	87.86	100/5
6	–	1451.47	72.57	–	100/5
7	10.5	1451.47	72.57	175.72	200/5
8	10.5	12906.89	645.34	52.71	700/5

Relay 3: $I_{nom3} = \dfrac{P_{nom3}}{\sqrt{3}V_3} = \dfrac{5.25 \times 10^6 \text{ VA}}{\sqrt{3} \times 13.2 \times 10^3 \text{ V}} = 229.63 \text{ A referred to } 13.2 \text{ kV}$

Relay 4: $I_{nom4} = \dfrac{P_{nom4}}{\sqrt{3}V_4} = \dfrac{5.25 \times 10^6 \text{ VA}}{\sqrt{3} \times 34.5 \times 10^3 \text{ V}} = 87.86 \text{ A referred to } 34.5 \text{ kV}$

Relay 5: $I_{nom5} = I_{nom4} = 87.86 \text{ A referred to } 34.5 \text{ kV}$

Relay 7: $I_{nom7} = \dfrac{P_{nom7}}{\sqrt{3}V_7} = \dfrac{10.5 \times 10^6 \text{ VA}}{\sqrt{3} \times 34.5 \times 10^3 \text{ V}} = 175.72 \text{ A referred to } 34.5 \text{ kV}$

Relay 8: $I_{nom8} = \dfrac{P_{nom8}}{\sqrt{3}V_8} = \dfrac{10.5 \times 10^6 \text{ VA}}{\sqrt{3} \times 115 \times 10^3 \text{ V}} = 52.71 \text{ A referred to } 115 \text{ kV}$

The short-circuit and load currents plus the selected CT ratio in line with the above-mentioned criteria are summarised in Table A.4.

Instantaneous, pick-up and time dial settings

Calculation of instantaneous settings

Relays 1 and 2:
$I_{inst} = 0.5 \times I_{sc} = 0.5 \times 1438.18 = 719.09$ primary amps.
Secondary amps $= 719.09 \times 5/150 = 23.97$ A.
Set relay to 24 secondary A, equivalent to 720 primary amps.

Relay 3: The instantaneous element is overridden to avoid the possibility of loss of discrimination with the transformer secondary protection.

Relay 4: The setting is based on 125 per cent of the short-circuit current which exists at the busbar on the lower voltage side of the transformer, referred to the higher voltage side.
$I_{inst} = 1.25 \times I_{sc}$, referred to $34.5 \text{ kV} = 1.25 \times 1438.18 \times (13.2/34.5) = 687.83$ primary amps, $\Rightarrow 687.83 \times 5/100 = 34.39$ secondary amps. Set at 35 secondary amps, equivalent to 700 primary amps.

Relay 5: The setting is calculated on the basis of 125 per cent of the current for the maximum fault level which exists at the next substation.
$I_{inst} = 1.25 \times I_{sc(B)} = 1.25 \times 881.43 = 1101.79$ primary amps, $\Rightarrow 1101.79 \times 5/100 = 55.09$ secondary amps. Set at 56 secondary amps, equivalent to 1120 primary amps.

Relay 6: $I_{inst} = 1000$ primary amps $= 50$ secondary amps.

Relay 7: The instantaneous element is overridden to avoid the possibility of loss of discrimination with the transformer secondary protection.

Relay 8: The setting is calculated on the basis of 125 per cent of the current for the maximum fault level which exists at the next substation.
$I_{inst} = 1.25 \times I_{sc(C)}$, referred to the higher voltage side, $= 1.25 \times 1451.47 \times (34.5/115) = 554.3$ primary amps, $\Rightarrow 554.3 \times 5/700 = 3.88$ secondary amps.
Set at 6 secondary amps (the minimum setting), equivalent to 840 primary amps.

Calculation of pick-up settings

Pick-up setting $= OLF \times I_{nom} \times (1/CTR)$
With an overload factor of 1.5:
Relays 1 & 2: $PU_{1,2} = 1.5 \times 114.81 \times (5/150) = 5.74$; set at 6
Relay 3: $PU_3 = 1.5 \times 229.63 \times (5/250) = 6.89$; set at 7
Relay 4: $PU_4 = 1.5 \times 87.86 \times (5/100) = 6.59$; set at 7
Relay 5: $PU_5 = 1.5 \times 87.86 \times (5/100) = 6.59$; set at 7
Relay 6: $PU_6 = 7$
Relay 7: $PU_7 = 1.5 \times 175.72 \times (5/200) = 6.59$; set at 7
Relay 8: $PU_8 = 1.5 \times 52.71 \times (5/700) = 056$; set at 1

Calculation of time dial settings

For the relays associated with breakers 1 and 2, choose the smaller dial setting, and calculate the operating time for the greater value of current that causes operation of the instantaneous unit.

Relays 1 and 2:

$$\text{Time dial setting} = 1/2.\ PSM = I_{inst.sec} \times (1/PU_{1,2}) = 24 \times (1/6) = 4 \text{ times}$$

With PSM of 4, and time dial setting of 1/2, from the operating characteristic of the relay, $t_1 = 0.16$ s.

Relay 3:
The relay backs up relays 1 and 2; therefore the required operating time t_3 is $0.16 + 0.4 = 0.56$ s.
PSM_3 is based on the 720 primary amps associated with relays 1 and 2, so that:

$$PSM_{3a} = 720 \times (1/CTR_3) \times (1/PU_3) = 720 \times (5/250) \times (1/7) = 2.06 \text{ times.}$$

With $PSM_{3a} = 2.06$, and a back-up operating time of 0.56 s, from the characteristic curve of a typical relay, time dial setting $= 1/2$.
    With this dial setting, relay 3 in fact acts as back-up to relay 1 and 2 in a time of 0.7 s. Since, in this case, the instantaneous unit is overridden the short-circuit current

is used and multiplied by the factor 0.86 in order to cover the delta-star transformer arrangement.

Thus, $PSM_{3b} = 0.86 \times I_{sc(A)} \times (1/CTR_3) \times (1/PU_3) = 0.86 \times 1438.18 \times (5/250) \times (1/7) = 3.53$ times.

With a time dial setting of 1/2, and $PSM_{3b} = 3.53$, the operating time of the relay, from the characteristic curve, is $t_3 = 0.6$ s.

Relay 4:
The operating time is based on this relay acting as back up to relay 3, i.e. $t_3 = 0.2 + 0.4 = 0.6$ s.
PSM_{4a} is based on the 1438.18 primary amps of the CT associated with relay 3, so that:

$$PSM_{4a} = 1438.18 \times (13.2\,kV/34.5\,kV) \times (5/100) \times (1/7) = 3.93 \text{ times.}$$

With $PSM_{4a} = 3.93$, and the back-up time of the relay $= 0.6$ s, this gives a time dial setting of 2. With this dial setting, relay 4 actually backs up relay 3 in a time of 0.78 s.

It is now necessary to calculate the operating time of relay 4 just before the operation of the instantaneous unit:

$$PSM_{4b} = I_{inst.prim.4} \times (1/CTR_4) \times (1/PU_4) = 700 \times (5/100) \times (1/7) = 5 \text{ times.}$$

With a time dial setting of 2, and $PSM = 5$, $t_4 = 0.49$ s.

Relay 5:
The required operating time is calculated on backing up relay 4, i.e. $t_5 = 0.49 + 0.4 = 0.89$ s. PSM_5 is based on the 700 primary amps of the CT associated with relay 4, so that, $PSM_5 = 700 \times (1/CTR_5) \times (1/PU_5) = 700 \times (5/100) \times (1/7) = 5$ times. With $PSM_5 = 5$ and a back-up time of 0.89 s, from the relay operating curve the time dial setting $= 4$.

With this dial setting, relay 5 will back up relay 4 in a time of one second.

It is now necessary to calculate the operating time of relay 5 just before the operation of the instantaneous unit:

$$PSM_{5a} = I_{inst.prim.5} \times (1/CTR_5) \times (1/PU_5) = 1120 \times (5/100) \times (1/7) = 8 \text{ times.}$$

With a time dial setting of 4 and $PSM = 8$, $t_5 = 0.45$ s.

Relay 6: time dial setting $= 5$

Relay 7:
Relay 7 backs up the operation of relays 5 and 6 and should be co-ordinated with the slower of these two relays.
Relay 6 has an instantaneous unit of 1000 primary amps, smaller than that for relay 5 which is set at 1120 amperes. Therefore, the operating time of both relays should be calculated for this value of current.

Relay 6: $PSM_6 = 1000 \times (5/100) \times (1/7) = 7.14$ times
With $PSM_6 = 7.14$ times and time dial setting $= 5$, $t_6 = 0.7$ s.
Relay 5: $PSM_7 = 1000 \times (5/100) \times (1/7) = 7.14$ times
With $PSM_5 = 7.14$ times and a time dial setting $= 4$, $t_5 = 0.52$ s.

Table A.5   Summary of settings for Exercise 5.3

Relay number	CT ratio	Pick-up (A)	Time dial	Instantaneous	
				primary (A)	secondary (A)
1 and 2	150/5	6	1/2	720	24
3	250/5	7	1/2	overridden	overridden
4	100/5	7	2	700	35
5	100/5	7	4	1120	56
6	100/5	7	5	1000	50
7	200/5	7	3	overridden	overridden
8	700/5	1	2	840	6

Therefore, in order to be slower than relay 6, the back-up time should be $t_7 = 0.7 + 0.4 = 1.1$ s.

It is now necessary to calculate the PSM that represents, in relay 7, the primary current of 1000 amps in the CT associated with relay 6, the relay with which it needs to be co-ordinated.

$$PSM_7 = 1000 \times (1/CTR_7) \times (1/PU_7) = 1000 \times (5/200) \times (1/7) = 3.57 \text{ times.}$$

With $t_7 = 1.1$ s and a time dial setting of 3.57, from the relay operating curve, time dial setting $= 3$.

As the instantaneous unit is overridden, the multiplier is calculated using the current for a short-circuit on busbar C:

$$PSM_7 = I_{sc7} \times (1/CTR_7) \times (1/PU_7) = 1451.47 \times (5/200) \times (1/7) = 5.18 \text{ times.}$$

With $PSM_7 = 5.18$ times, and a time dial setting $= 3$, $t_7 = 0.7$ s.

Relay 8:
The back-up time of relay 8 over relay 7 is given by $t_8 = 0.7 + 0.4 = 1.1$ s.
Now calculate the multiplier that represents in relay 8 the 1451.47 primary amps referred to 115 kV:

$$MULT_8 = 1451.47 \times (34.5 \text{ kV}/115 \text{ kV}) \times (5/700) \times 1 = 3.11 \text{ times.}$$

With $MULT_8 = 3.11$ times, and $t_8 = 1.1$ seconds, from the relay curve $DIAL_8 = 2$.

Coverage of protection

$$\% \text{ cover} = \frac{K_s(1 - K_i) + 1}{K_i},$$

where

$$K_i = \frac{I_{sc.pickup}}{I_{sc.end}} = \frac{I_{inst.prim.5}}{I_{sc(B)}} = \frac{1120 \text{ primary A}}{881.43 \text{ A}} = 1.2707$$

and

$$K_s = \frac{Z_{source} + Z_{T1}}{Z_{line}} = \frac{0.463 + 13.26}{8.875} = 1.5463$$

$$\% \text{ cover} = \frac{1.5463(1 - 1.2707) + 1}{1.2707} = 0.4576$$

Therefore the cover by the instantaneous unit of the overcurrent relay associated with breaker 5 is 45.7 per cent of the 34.5 kV line.

Solution to Exercise 7.1

Figure A.6 shows the network for Exercise 7.1.

Calculation of short-circuit levels

Busbar A: $I_{sc} = 6560$ A

Busbar B:
(a) With the ring closed the equivalent network is as shown in Figure A.7.

$$I_{sc(B)} = \frac{13.2 \times 10^3}{\sqrt{3} \times (1.16 + 1.556)} = 2806.39 \text{ A}$$

Taking account of the division of current in the network, the fault current that circulates through breaker 1 is $2806.39 \times (7/9) = 2182.75$ A, and the current that circulates through breaker 2 is $2806.39 \times (2/9) = 623.64$ A.
(b) With the ring open.

With breaker 1 open, $I_{sc(B)} = \dfrac{13.2 \times 10^3}{\sqrt{3} \times (1.16 + 7)} = 933.95 \text{ A} = I_{sc.max}$ for relay 3

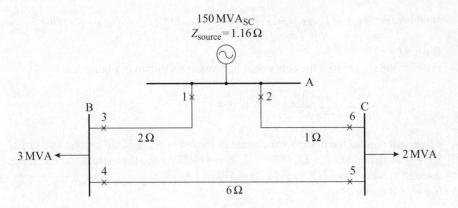

Figure A.6   Network for Exercise 7.1

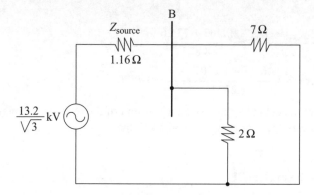

Figure A.7   Equivalent network for fault on busbar B

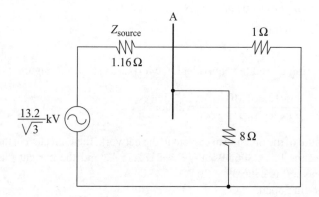

Figure A.8   Equivalent network for fault on busbar C

With breaker 2 open, $I_{\mathrm{sc(B)}} = \dfrac{13.2 \times 10^3}{\sqrt{3} \times (1.16+2)} = 2411.72\,\mathrm{A} = I_{\mathrm{sc.max}}$ for relay 4

Busbar C:

(a) With the ring closed the equivalent network is as shown in Figure A.8.

$$I_{\mathrm{sc(C)}} = \frac{13.2 \times 10^3}{\sqrt{3} \times (1.16+0.889)} = 3719.39\,\mathrm{A}$$

Taking account of the division of current in the network, the fault current that circulates through breaker 1 is $3719.39 \times (1/9) = 413.27\,\mathrm{A}$, and the current that circulates through breaker 2 is $3719.39 \times (8/9) = 3306.12\,\mathrm{A}$.

(b) With the ring open.

With breaker 1 open, $I_{\mathrm{sc(C)}} = \dfrac{13.2 \times 10^3}{\sqrt{3} \times (1.16+1)} = 3528.25\,\mathrm{A} = I_{\mathrm{sc.max}}$ for relay 5

With breaker 2 open, $I_{sc(C)} = \dfrac{13.2 \times 10^3}{\sqrt{3} \times (1.16 + 8)} = 831.99\,\text{A} = I_{sc.max}$ for relay 6

Calculation of CT ratios

First, calculate the maximum nominal currents for each breaker.
Breaker 1 open:

$$I_{max,2\&6} = \frac{5 \times 10^6\,\text{VA}}{\sqrt{3} \times (13.2 \times 10^3\,\text{V})} = 218.69\,\text{A}.$$

$$I_{max,4\&5} = \frac{3 \times 10^6\,\text{VA}}{\sqrt{3} \times (13.2 \times 10^3\,\text{V})} = 131.22\,\text{A}.$$

$$I_{max,1\&3} = 0$$

Breaker 2 open:

$$I_{max,1\&3} = \frac{5 \times 10^6\,\text{VA}}{\sqrt{3} \times (13.2 \times 10^3\,\text{V})} = 218.69\,\text{A}.$$

$$I_{max,4\&5} = \frac{2 \times 10^6\,\text{VA}}{\sqrt{3} \times (13.2 \times 10^3\,\text{V})} = 87.48\,\text{A}.$$

$$I_{max,2\&6} = 0$$

The CTs are selected, starting from the maximum nominal currents for each breaker, and checking the saturation of maximum fault current at each breaker at $0.05 \times I_{sc}$:

Breaker 1 : $I_{max\,1} = 218.69 \Rightarrow CT_1$ ratio $= 250/5$
Breaker 4 : $I_{max\,4} = 131.22 \Rightarrow CT_4$ ratio $= 200/5$
Breaker 6 : $I_{max\,6} = 218.69 \Rightarrow CT_6$ ratio $= 250/5$

Setting of instantaneous units

In order to set relays 1, 4 and 6, the ring is opened at breaker 2 and the setting for each one of the relays is calculated using the maximum short-circuit current that is seen by the relay which is backing up.

Relay 1: this relay backs up relay 4.
$I_{inst} = 1.25 \times I_{sc.4} = 1.25 \times 2411.72\,\text{A} = 3014.65$ primary amps, i.e. $3014.65 \times (5/250) = 60.29$ secondary amps. Set to 61 secondary amps, equivalent to 3050 primary amps.

Relay 4: this relay backs up relay 6.
$I_{inst} = 1.25 \times I_{sc.6} = 1.25 \times 831.99\,\text{A} = 1040$ primary amps, i.e. $1040 \times (5/200) = 26$ secondary amps. Set to 26 secondary amps, equivalent to 1040 primary amps.

Relay 6:
Set the relay taking 125 per cent of the maximum load current, which comes from the source: $I_{inst} = 1.25 \times I_{max\,.6} = 1.25 \times 218.69\,\text{A} = 273.36$ primary amps,

i.e. $273.36 \times (5/250) = 5.47$ secondary amps. Set to 6 secondary amps, equivalent to 300 primary amps.

Calculation of pick-up settings

Pick-up setting $= \text{OLF} \times I_{nom} \times (1/\text{CTR})$
Taking an overload factor of 1.5
Breaker 2 open:
Relay 1: $\text{PU}_1 = 1.5 \times I_{nom1} \times (1/\text{CTR}_1) = 1.5 \times 218.69 \times (5/250) = 6.56$;
  set $\text{PU}_1 = 7$
Relay 4: $\text{PU}_4 = 1.5 \times I_{nom4} \times (1/\text{CTR}_4) = 1.5 \times 131.22 \times (5/200) = 4.92$;
  set $\text{PU}_4 = 5$
Relay 6: $\text{PU}_6 = 1.5 \times I_{nom6} \times (1/\text{CTR}_6) = 1.5 \times 218.69 \times (5/250) = 6.56$;
  set $\text{PU}_6 = 7$

Calculation of time dial settings

Here the IEC expression, given in eqn. 5.7, Section 5.3.3 in Chapter 5, is used when considering very inverse time relays.

Relay 6:
The setting is the lowest time dial setting possible, and for this type of relay this is a setting of 0.1.

$$\text{PSM}_6 = 831.99 \times (1/\text{CTR}_6) \times (1/\text{PU}_6) = 831.99 \times (5/250) \times (1/7) = 2.38.$$

With a time dial setting of 0.1 and $\text{PSM}_6 = 2.38$, using the expression for a very inverse relay where $\alpha = 1$ and $\beta = 13.5$:

$$t_6 = \frac{0.1 \times 13.5}{2.38 - 1} = 0.98 \, s$$

Relay 4:
The back-up time of relay 4 over relay 6, $t_4 = 0.98 + 0.4 = 1.38 \, s$.
The multiplier that represents the 831.99 A of relay 6 is calculated for relay 4:

$$\text{PSM}_4 = 831.99 \times (1/\text{CTR}_4) \times (1/\text{PU}_4) = 831.99 \times (5/200) \times (1/5) = 4.16.$$

With $t_4 = 1.38 \, s$ and $\text{PSM}_4 = 4.16$, using the expression for a very inverse relay where $\alpha = 13.5$ and $\beta = 1$:

$$t_4 \Rightarrow 1.38 = \frac{\text{time dial setting} \times 13.5}{4.16 - 1} \Rightarrow \text{time dial setting} = 0.32.$$

The time dial setting chosen is 0.4.
Next calculate the multiplier that represents $I_{sc.max}$ for relay 4 (2411.72 A):

$$\text{PSM}_{4a} = 2411.72 \times (1/\text{CTR}_4) \times (1/\text{PU}_4) = 2411.72 \times (5/200) \times (1/5) = 12.06.$$

With a time dial setting $= 0.4$ and $\text{PSM}_{4a} = 12.06$, using the expression for a very inverse relay where $\alpha = 1$ and $\beta = 13.5$:

$$t_4 = \frac{\text{time dial setting} \times 13.5}{\text{PSM}_{4a} - 1} = \frac{0.4 \times 13.5}{12.06 - 1} = 0.49 \, s.$$

Relay 1:
The back-up time of relay 1 over relay 4, t_1, $= 0.49 + 0.4 = 0.89$ s. The multiplier that represents the 2411.72 A of relay 4, at relay 1, is calculated as:

$$PSM_1 = 2411.72 \times (1/CTR_1) \times (1/PU_1) = 2411.72 \times (5/250) \times (1/7) = 6.89.$$

With $t_1 = 0.89$ s and $PSM_1 = 6.89$, using the expression for a very inverse relay:

$$t_1 = \frac{\text{time dial setting} \times 13.5}{PSM_1 - 1} \Rightarrow 0.89$$

$$= \frac{\text{time dial setting} \times 13.5}{6.89 - 1} \Rightarrow \text{time dial setting} = 0.39 \text{ s}.$$

The time dial setting chosen is 0.4.

Solution to Exercise 7.2

The network is shown in Figure A.9.

Calculation of short-circuit currents

Busbar B:
With the ring closed, $I_{sc(B)} = 2160$ A, and with the ring open $I_{sc(B)} = 2880$ A, the largest short-circuit current for relay B.
Busbar C:
With the ring closed, $I_{sc(C)} = 1350$ A, and with the ring open $I_{sc(C)} = 2300$ A, the largest short-circuit current for relay C.

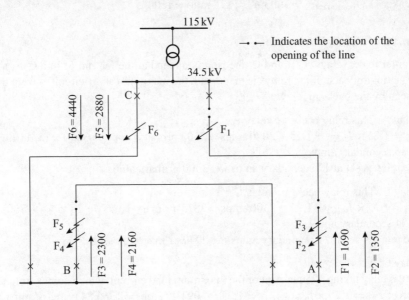

Figure A.9   Network for Exercise 7.2

Calculation of maximum nominal currents

These are calculated with the ring open at the far end of line C-A, and with the ring open at breaker A, in order to determine the larger value of current.

Ring open at the far end of line C-A:

$$I_{\text{nom.max.(A)}} = \frac{25 \times 10^6 \text{ VA}}{\sqrt{3} \times (13.2 \times 10^3 \text{ V})} = 1093.47 \text{ A}$$

$$I_{\text{nom.max.(B)}} = \frac{15 \times 10^6 \text{ VA}}{\sqrt{3} \times (13.2 \times 10^3 \text{ V})} = 656.1 \text{ A}$$

$$I_{\text{nom.max(C)}} = 0$$

Ring open at breaker A:

$$I_{\text{nom.max.(C)}} = \frac{25 \times 10^6 \text{ VA}}{\sqrt{3} \times (13.2 \times 10^3 \text{ V})} = 1093.47 \text{ A}$$

$$I_{\text{nom.max.(B)}} = \frac{10 \times 10^6 \text{ VA}}{\sqrt{3} \times (13.2 \times 10^3 \text{ V})} = 437.39 \text{ A}$$

$$I_{\text{nom.max(A)}} = 0$$

The CTs are selected, starting from the maximum nominal currents for each breaker, and checking the saturation of maximum fault current at each breaker at $0.05 \times I_{\text{sc}}$:

Breaker A : $I_{\text{nom.max(A)}} = 1093.47 \Rightarrow$ CT ratio $= 1100/5$
Breaker B : $I_{\text{nom.max(B)}} = 656.10 \Rightarrow$ CT ratio $= 700/5$
Breaker C : $I_{\text{nom.max(C)}} = 1093.47 \Rightarrow$ CT ratio $= 1100/5$

Setting of instantaneous units

In order to set relays A, B and C, the ring is opened at the far end of line C-A, and the setting for each of the relays is made with the maximum short-circuit current that is seen by the back-up relay.

Relay A: This relay backs up relay B
$I_{\text{inst}} = 1.25 \times I_{\text{sc(B)}} = 1.25 \times 2880 \text{ amps} = 3600 \text{ primary amps} \Rightarrow 3600 \times (5/1100) = 16.36$ secondary amps.
The relay is set at 17 secondary amps i.e. 3300 primary amps.

Relay B: This relay backs up relay C
$I_{\text{inst}} = 1.25 \times I_{\text{sc(C)}} = 1.25 \times 2300 \text{ amps} = 2875 \text{ primary amps} \Rightarrow 2875 \times (5/700) = 20.54$ secondary amps.
The relay is set at 21 secondary amps i.e. 2940 primary amps.

Relay C:
Set the relay taking 125 per cent of the maximum load current, which comes from the source: $I_{\text{inst}} = 1.25 \times I_{\text{nom.max(C)}} = 1.25 \times 1093.47 \text{ amps} = 1366.84 \text{ primary amps} \Rightarrow 1366.84 \times (5/1100) = 6.21$ secondary amps.

The relay is set at 7 secondary amps i.e. 1540 primary amps.
The pick-up of each relay is 5.

Time dial setting

The settings of relays A, B and C are determined with the ring open at the far end of line C-A.

Relay C:
This relay is set using the smallest dial setting, i.e. 1/2.

$$PSM_C = I_{sc.max(C)} \times (1/CTR_C) \times (1/PU_C) = 2300 \times (5/1100) \times (1/8) = 1.3$$

With a time dial setting $= 1/2$ and $PSM_C = 1.3$, from the relay curves, $t_C = 2$ s.

Relay B:
The back-up time of relay B over relay C, $t_B, = 2 + 0.4 = 2.4$ s.
The multiplier that represents the 2300 A of relay C is calculated for relay B:

$$PSM_{B1} = 2300 \times (1/CTR_B) \times (1/PU_B) = 2300 \times (5/700) \times (1/8) = 2.1$$

With $t_B = 2.4$ s, and $PSM_{B1} = 2.1$, from the relay curves, time dial setting $= 2$.
    Next calculate the multiplier which represents $I_{sc.max}$ for relay B (2880 A):

$$PSM_{B2} = I_{sc.max(B)} \times (1/CTR_B) \times (1/PU_B) = 2880 \times (5/700) \times (1/8) = 2.6$$

With time dial setting $= 2$, and $PSM_{B2} = 2.6$, from the relay curves, $t_B = 1.8$ s.

Relay A:
The back-up time of relay A over relay B, $t_A, = 1.8 + 0.4 = 2.2$ s.
The multiplier that represents the 2880 A of relay B is calculated for relay A:

$$PSM_A = I_{sc.max(A)} \times (1/CTR_A) \times (1/PU_A) = 2880 \times (5/1100) \times (1/8) = 1.7$$

With $t_A = 2.2$ s and $PSM_A = 1.7$, from the relay curves, time dial setting $= 1$.

Solution to Exercise 8.1

The diagram for this exercise is shown in Figure A.10.

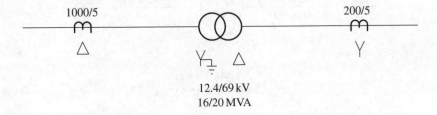

12.4/69 kV
16/20 MVA

Figure A.10   Diagram for Exercise 8.1

The nominal currents for maximum load on the transformer are calculated as

$$I_{nom(69kV)} = \frac{20 \times 10^6 \, MVA}{\sqrt{3} \times 69 \times 10^3 \, V} = 167.35 \, A$$

and

$$I_{nom(12.4kV)} = \frac{20 \times 10^6 \, MVA}{\sqrt{3} \times 12.4 \times 10^3 \, V} = 931.21 \, A$$

The currents in the relays are

$$I_{relay(12.4kV)} = 931.21 \times (5/1000) \times \sqrt{3} = 8.06 \, A; \quad PU = 8.7 \text{ is therefore chosen.}$$

$$I_{relay(69kV)} = 167.35 \times (5/200) \times \sqrt{3} = 4.18 \, A; \quad PU = 4.2 \text{ is therefore chosen.}$$

Solution to Exercise 8.2

The diagram for this exercise is shown in Figure A.11.

The nominal currents for maximum load on the transformer are calculated as

$$I_{nom(161kV)} = \frac{40 \times 10^6 \, MVA}{\sqrt{3} \times 161 \times 10^3 \, V} = 143.44 \, A$$

$$I_{nom(69kV)} = \frac{40 \times 10^6 \, MVA}{\sqrt{3} \times 69 \times 10^3 \, V} = 334.7 \, A$$

$$I_{nom(12.4kV)} = \frac{10 \times 10^6 \, MVA}{\sqrt{3} \times 12.4 \times 10^3 \, V} = 456.61 \, A$$

and the currents in the relays are therefore

$$I_{relay(161kV)} = 143.44 \times (5/200) \times \sqrt{3} = 6.2 \, A; \quad PU \text{ of } 6.2 \text{ is chosen.}$$

$$I_{relay(69kV)} = 334.7 \times (5/400) = 4.18 \, A; \quad PU \text{ of } 4.2 \text{ is chosen.}$$

$$I_{relay(12.4kV)} = 456.61 \times (5/500) = 4.65 \, A; \quad PU \text{ of } 4.6 \text{ is chosen.}$$

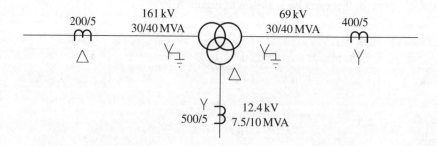

Figure A.11   Diagram for Exercise 8.2

Solution to Exercise 8.3

The diagram for this exercise is shown on Figure A.12.
The nominal currents under normal conditions are

$$I_{nom.34.5\,kV} = \frac{10 \times 10^6 \text{ VA}}{\sqrt{3} \times 34.5 \times 10^3 \text{ V}} = 167.35 \text{ A}$$

$$I_{nom.13.2\,kV} = \frac{10 \times 10^6 \text{ VA}}{\sqrt{3} \times 13.2 \times 10^3 \text{ V}} = 437.39 \text{ A}$$

The vector diagram for the Dy1 connection is shown in Figure A.13, taking into account that the rotation of the phases A, B and C is negative, i.e. clockwise.

The diagram showing the flow of currents in the primary and secondary windings of the transformer, and in the CT secondaries is given in Figure A.14.

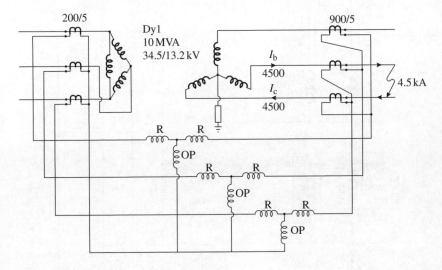

Figure A.12   Diagram for Exercise 8.3

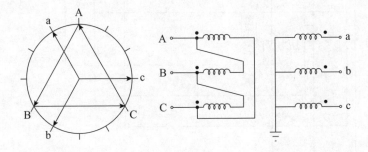

Figure A.13   Vector group for Dy1 transformers

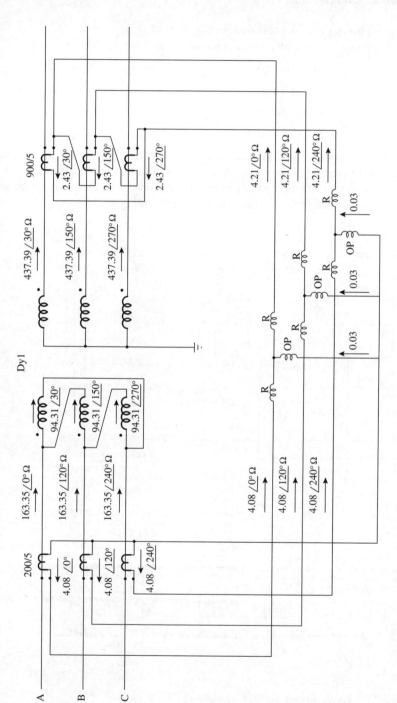

Figure A.14   Diagram for normal conditions (Exercise 8.3)

For a fault between phases B and C, the current that is induced in the delta is

$$I_{\text{Delta}} = I_{\text{fault}} \times \frac{N_2}{N_1},$$

where

$$\frac{N_2}{N_1} = \frac{V_2}{\sqrt{3} \times V_1}$$

Therefore

$$I_{\text{Delta}} = I_{\text{fault}} \times \frac{V_2}{\sqrt{3} \times V_1} = 4500 \times \frac{13.2}{\sqrt{3} \times 34.5} = 994.05 \, \text{A}$$

The line currents are equal to those in the delta for two of the phases and double the current for the phase that has the larger current.

The schematic diagram in Figure A.15 shows the fault between phases B and C and the different magnitudes and directions of the currents in the conductors and those that circulate due to the fault. Starting from the figure it is necessary to determine if the differential relays would operate, taking into account whether any unbalance exists in the currents through the operating coils.

As will be seen from Figure A.15, only a small amount of current flows through the operating coils of the differential relays, because there is no large unbalance of currents due to the fault, and therefore the differential relays do not operate for a fault between conductors B and C even though the phase rotation is negative.

Solution to Exercise 8.4

Figure A.16 shows the three-phase primary and secondary connections for Exercise 8.4, while Figure A.17 provides the single-line diagram.

The maximum load currents of the transformer are given by

$$I_{\text{nom}(69\,\text{kV})} = \frac{20 \times 10^6 \, \text{VA}}{\sqrt{3} \times 69 \times 10^3 \, \text{V}} = 167.35 \, \text{A}$$

$$I_{\text{nom}(12.4\,\text{kV})} = \frac{20 \times 10^6 \, \text{VA}}{\sqrt{3} \times 12.4 \times 10^3 \, \text{V}} = 931.21 \, \text{A}$$

and the currents in the relays are

$$I_{69\,\text{kV}} = 167.35 \times (5/600) = 1.39 \, \text{A}$$

$$I_{12.4\,\text{kV}} = 931.21 \times (5/1200) \times (1/a) \times (1/\sqrt{3})$$

Since the currents that enter the relay from each side must balance, the transformation ratio of the compensation transformer is obtained from $931.21 \times (5/1200) \times (1/a) \times (1/\sqrt{3}) = 1.39 \, \text{A}$, from which $a = 1.61$.

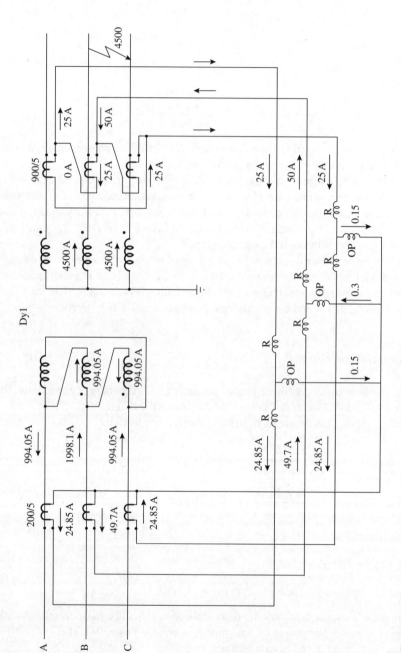

Figure A.15　Diagram for fault between phases B and C (Exercise 8.3)

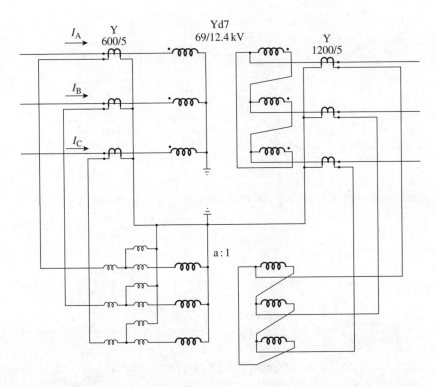

Figure A.16   Diagram for Exercise 8.4

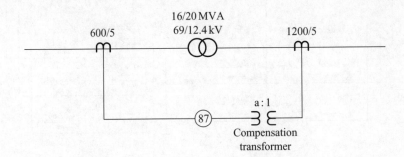

Figure A.17   Single-line diagram for Exercise 8.4

The phasor diagram for the Yd1 connection is given in Figure A.18, taking into account that the rotation of the phases A, B and C is positive.

From the phase diagram for a Yd7 connection, the three-phase connection diagram in Figure A.19 is obtained, showing the connection of the CTs, and the magnitude and direction of the currents in the primary and secondary connections.

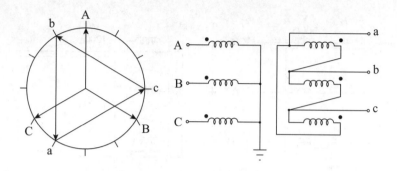

Figure A.18   Vector group for Yd7 transformers

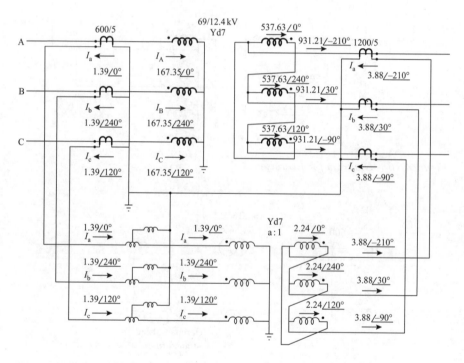

Figure A.19   Diagram for normal conditions for Exercise 8.4

Solution to Exercise 9.1

The characteristic of a mho relay in a reactance diagram is a circle that passes through the origin of the axis of the co-ordinates, and with a centre at the point (a, b), as in Figure A.20.

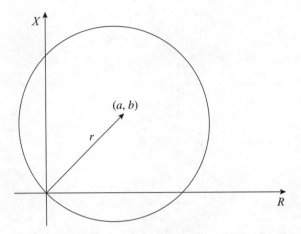

Figure A.20   Mho characteristics in X-R diagram (Exercise 9.1)

In order to obtain the characteristic of the mho relay in the admittance diagram it is necessary to create the characteristic of the relay as a function of these values:

$$Z = R + jX$$

$$Y = 1/(R + jX)$$

Multiplying both sides by $R - jX$ gives

$$Y = (R - jX)/(R^2 + X^2) = G + jB$$

Therefore

$$G = R/(R^2 + X^2)$$

and

$$B = -X/(R^2 + X^2)$$

In accordance with the equation for a circle:

$$(R - a)^2 + (X - b)^2 = r^2$$

$$R^2 - 2aR + a^2 + X^2 - 2bX + b^2 = r^2$$

where $r^2 = a^2 + b^2$
Thus:

$$R^2 - 2aR + X^2 - 2bX = 0$$

Dividing the above expression by $(R^2 + X^2)$ gives

$$\frac{R^2 + X^2}{R^2 + X^2} - \frac{2aR}{R^2 + X^2} - \frac{2bX}{R^2 + X^2} = 0$$

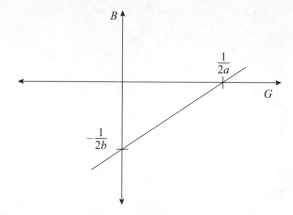

Figure A.21   Mho characteristics in B-G diagram (Exercise 9.1)

and

$$1 - \frac{2aR}{R^2 + X^2} - \frac{2bX}{R^2 + X^2} = 0$$

Replacing G and B in the expression gives:

$$1 - 2aG + 2bB = 0$$

$$B = \frac{aG}{b} - \frac{1}{2b}$$

If $B = 0$, then $G = 1/2a$; and if $G = 0$, then $B = -1/2b$, so that the characteristic of the mho relay in the admittance diagram will be as shown in Figure A.21.

Solution to Exercise 9.2

The diagram for exercise 9.2 is shown in Figure A.22.
For a single-phase fault the three sequence networks are connected in series as shown in Figure A.23.
For a single-phase fault it can be shown that

$$I_{a1} = I_{a2} = I_{a0} \Rightarrow I_{a1} = I_a/3$$

$$I_a = 3I_{a1} = \frac{3(11.8 \times 10^3)}{\sqrt{3}\{2 \times (j0.23 + 2.3 + j5.7) + 45 + j7.5 + (3.5 + j24) + 3R_F\}}$$

$$I_a = \frac{(20.44 \times 10^3)}{(53.1 + 3R_F + j43.46)}$$

If R_T is taken as equal to $53.1 + 3R_F$, then $R_T + j43.36 = (20.44 \times 10^3)/(I_F \angle -\theta_1)$.
    If $I_F = 200$ amps, then $R_T + j43.36 = (20.44 \times 10^3)/(200\angle -\theta_1)$, and

$$R_T = \sqrt{102.2^2 + 43.36^2} = 92.25 \, \Omega.$$

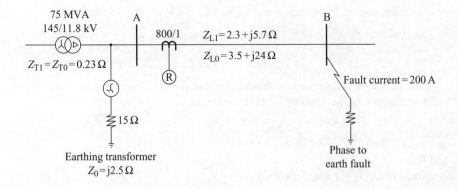

Figure A.22   Diagram for Exercise 9.2

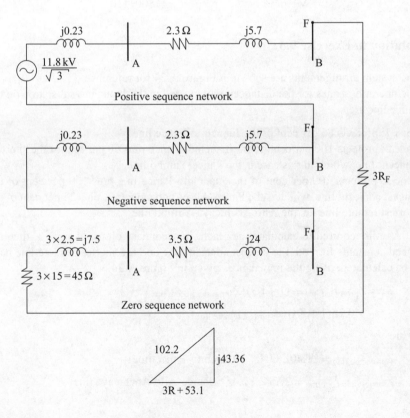

Figure A.23   Sequence networks for Exercise 9.2

Therefore the fault resistance, $R_F, = (92.55 - 53.1)/3 = 13.15\,\Omega$.

The residual compensation constant, K, is given by $K = \dfrac{Z_{L0} - Z_{L1}}{3 \times Z_{L1}}$

$K = (3.5 + j24 - 2.3 - j5.7)/3(2.3 + j5.7) = (1.2 + j18.3)/(6.9 + j17.1) = 0.99$

Calculation of the secondary impedance that the relay sees if it is used with a residual compensation equal to 1.0 (100 per cent):

$$Z_R = \frac{V_{Ra}}{I_{Ra}} = \frac{I_0 \times (Z_{L1} + Z_{L2} + 3R_F + Z_{L0})}{3I_0 + KI_R} = \frac{2Z_{L1} + 3R_F + Z_{L0}}{3(1 + K)},$$

and $K = 1$.

Since $I_R = 3I_0$ for this fault, then

$$Z_R = \frac{2(2.3 + j5.7) + (3 \times 13.15) + (3.5 + j24)}{6} = 7.93 + j5.9$$

The secondary ohms are

$$\text{primary ohms} \times \frac{\text{CT ratio}}{\text{VT ratio}} = (7.93 + j5.9) \times \frac{800/1}{11800/110} = 59.14 + j44\,\Omega$$

Solution to Exercise 9.3

The system arrangements are shown in Figure A.24 for reference.

The operating zones for the distance relay located in the Juanchito substation on the Pance line are

Zone 1: protects 85 per cent of the Juanchito-Pance line.
Zone 2: protects 100 per cent of the Juanchito-Pance plus 50 per cent of the shortest adjacent line, which in this case is the Pance-Yumbo line.
Zone 3: protects 100 per cent of the Juanchito-Pance line plus 100 per cent of the longest adjacent line, which is the Pance-Alto Anchicayá line, plus 25 per cent of the shortest remote line i.e. the Alto Anchicayá-Yumbo line.

As it is required to calculate the reach, in secondary ohms, of zone 3, then the infeed constants for both the longest adjacent line and the shortest remote line need to be calculated, using the impedances given in Figure A.24.

$$Z_3 = Z_{\text{protected line}} + (1 + K_2)(Z_{\text{longest adjacent line}})$$
$$+ 0.25(1 + K_3)(Z_{\text{shortest remote line}}),$$

where

$$Z_{\text{protected line}} = 11.40\angle 83.48° \text{ (Juanchito-Pance line)}$$

$$Z_{\text{longest adjacent line}} = 27.64\angle 82.45° \text{ (Pance-Alto Anchicayá line)}$$

and

$$Z_{\text{shortest remote line}} = 27.84\angle 82.45° \text{(Alto Anchicayá-Yumbo line)}$$

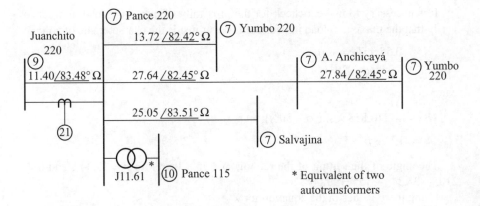

Figure A.24   Power system for Exercise 9.3

Thus:

$$K_2 = \frac{I_{\text{not seen by relay}}}{I_{\text{relay}}} = \frac{I_{2-7} + I_{10-7} + I_{8-7} + 0.5(I_{9-7})}{0.5(I_{9-7})}$$

From the printout in Figure 9.53:

$$K_2 = \frac{1278\angle - 87.96° + 1210.67\angle - 87.6° + 2446.77\angle - 86.18° + 0.5(1814.2\angle - 86.59°)}{0.5(1814.2\angle - 86.59°)}$$

$$= 6.44\angle - 0.34°$$

The calculation of the second infeed constant for the coverage of the shortest remote line, K_3, is similar to that for the first infeed constant for the longest adjacent line K_2, (calculated above), but it is now necessary to take into account the contribution of the generation at Alto Anchicayá. This contribution does not affect the coverage of the Zone 3 relay on the Pance-Alto Anchicayá line, but does on the remote Alto Anchicayá-Yumbo circuit. Therefore

$$K_3 = \frac{I_{\text{not seen by relay}}}{I_{\text{relay}}} = \frac{I_{2-7} + I_{10-7} + I_{8-7} + 0.5(I_{9-7}) + I_{\text{contribution from gen at AA}}}{0.5(I_{9-7})}$$

The current contributions from the Alto Anchicayá substation that should be included in the above formula are not available for this exercise. However, K_3 can be assumed to be equal to K_2, so that the reach of Z_3 will be reduced by a small amount, which is not critical since this zone acts as a remote back-up as mentioned in Section 9.3.

$$Z_3 = 11.4\angle 83.5° + \{(1 + 6.44\angle - 0.34°) \times 27.64\angle 82.45°\}$$
$$+ \{0.25(1 + 6.44\angle - 0.34°) \times 27.84\angle 82.45°\}$$
$$= 268.82\angle 82.21° \text{ primary ohms}$$

The value of Z_3 in secondary ohms is

$$Z_3 = (\text{primary ohms}) \times (\text{CT ratio})/(\text{VT ratio})$$
$$= 268.82\angle 82.21° \times (800/5) \times (1/2000) = 21.5 \text{ secondary ohms}$$

It is necessary to make a check for the proximity of maximum load in order to verify that the maximum load impedance is never inside the characteristic of zone 3:

$$Z_x = \frac{0.55 \times Z_3 \times \sin\alpha}{\sin(\phi - 30°)}$$

where

$$\beta = \sin^{-1}\{0.818 \times \sin(\phi - 30°)\}$$
$$\alpha = 210° - \beta - \phi$$

The angle of the setting of the relay is $\phi = 75°$; therefore $\beta = 35.34°$, and $\alpha = 210° - 35.34° - 75° = 99.66°$.

Using these values in the equation for Z_x:

$$Z_x = \frac{0.55 \times 268.82 \times \sin 99.66°}{\sin(75° - 30°)} = 206.13 \text{ primary ohms.}$$

The maximum load impedance is

$$Z_c = \frac{V^2}{P} = \frac{(220 \times 10^3)^2}{40 \times 10^6} = 1210 \text{ primary ohms}$$

The check consists of verifying that

$$\frac{Z_c - Z_x}{Z_c} \geq 0.5$$

For the double circuit line:

$$\frac{Z_c - Z_x}{Z_c} = \frac{1210 - 206.13}{1210} = 0.829$$

which is greater than 0.5. Therefore the setting of the relay can be considered appropriate, and it is not necessary to reduce its reach.

Index